GUIDEBOOK FOR UTILITIES-LED BUSINESS MODELS

WAY FORWARD FOR ROOFTOP SOLAR IN INDIA

DECEMBER 2022

ASIAN DEVELOPMENT BANK

ADB

ISBN 978-92-9269-825-6 (print); 978-92-9269-826-3 (electronic); 978-92-9269-827-0 (ebook)
Publication Stock No. TIM220555-2
DOI: http://dx.doi.org/10.22617/TIM220555-2

The views expressed in this publication are those of the authors and do not necessarily reflect the views and policies of the Asian Development Bank (ADB) or its Board of Governors or the governments they represent.

ADB does not guarantee the accuracy of the data included in this publication and accepts no responsibility for any consequence of their use. The mention of specific companies or products of manufacturers does not imply that they are endorsed or recommended by ADB in preference to others of a similar nature that are not mentioned.

By making any designation of or reference to a particular territory or geographic area, or by using the term "country" in this document, ADB does not intend to make any judgments as to the legal or other status of any territory or area.

Corrigenda to ADB publications may be found at http://www.adb.org/publications/corrigenda.

Notes:
1. In this publication, Ministry of New and Renewable Energy wherever mentioned, refers to the said Union Ministry of the Government on India.
2. "FY" before a calendar year denotes the year in which the fiscal year ends, e.g., FY2022 ends on 31 March 2022.
3. ADB recognizes "Bangalore" as Bengaluru.

On the cover: Installing photovoltaic panels could transform any sun-kissed roof into an energy-harvesting system. Photo 1 (ID 213137836 © Yaman Kumar | Dreamstime.com) shows a construction worker laying energy-efficient slate tiles on the roof of a building in Mandi in the state of Himachal Pradesh in northern India. Photo 2 (ID 61976086 © Diianadimitrova | Dreamstime.com) shows photovoltaic panels.

CONTENTS

TABLES, FIGURES, AND BOXES

Tables

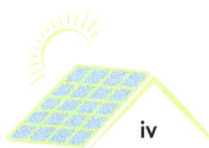

Figures

Boxes

ACKNOWLEDGMENTS

This publication *Guidebook for Utilities-Led Business Models: Way Forward for Rooftop Solar in India* was conceptualized and prepared under the leadership of Jigar Bhatt, energy specialist, Asian Development Bank (ADB), India Resident Mission as a part of Technical Assistance (TA) for facilitating Solar Rooftop Investment Program (SRIP) in India. This guidebook has been prepared by KPMG Advisory Services Private Limited and Idam Infrastructure Advisory Private Limited under the ADB SRIP TA. We are indebted to the ADB Management for providing valuable guidance, support, and encouragement for preparing this document. It would not have been possible to present this work in its current form without insightful review and excellent contribution from the agencies and peers—all of which are sincerely appreciated.

We would like to take this opportunity to thank Vandana Kumar, Additional Secretary and Amitesh Kumar Sinha, Former Joint Secretary, Ministry of New and Renewable Energy (MNRE), Government of India for motivating us to undertake this study. We would also like to express our sincere gratitude to Jeevan Kumar Jethani, Senior Director, MNRE for not only sharing his time and valuable knowledge but for supporting us through the process of developing and finalizing this publication.

We also take this opportunity to thank the managing director and other senior management of the Kerala State Electricity Board Limited (KSEBL) who have shown immense trust in the ADB SRIP TA Program and taken a lead in conceptualizing and implementing new and innovative utility-led business models by developing appropriate risk and reward frameworks for stakeholders. Lessons learned while supporting KSEBL in implementing these business models are included as a case study in this guidebook. We also thank the SUPRABHA TA Program of World Bank and Indo German Energy Program, GIZ for supporting us with their valuable comments and providing us access to best practices and success stories from the states supported by them.

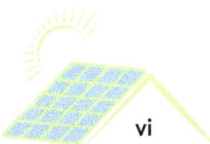

ABBREVIATIONS

ABR	average billing rate
APPC	average power purchase cost
CAPEX	capital expenditure
CUF	capacity utilization factor
DISCOM	distribution company
EPC	engineering, procurement, and construction
FIT	Feed-in tariff
GW	gigawatt
kWh	kilowatt-hour
kWp	kilowatt-peak
MNRE	Ministry of New and Renewable Energy
MW	megawatt
O&M	operation and maintenance
OPEX	operational expenditure
PPA	power purchase agreement
REC	renewable energy certificates
RESCO	Renewable Energy Service Company
RPO	renewable purchase obligation
RTS	rooftop solar

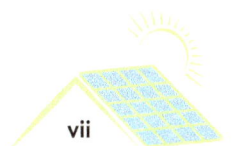

EXECUTIVE SUMMARY

India set a target: generating 175 gigawatts (GW) of renewable energy by 2022. While 100 GW is now available, 50 GW is being installed, and 27 GW is still subject to tender. The Government of India has in the meanwhile also raised its target, to install 500 GW of renewable energy capacity by 2030.

Within renewable energy, India had aimed to generate 100 GW of solar energy, and produce 40 GW of that output through rooftop solar (RTS) projects. Despite the enabling policy environment and attractive economics, progress on solar power generation has been slow. According to the physical progress reported by the Ministry of New and Renewable Energy (MNRE), Government of India, the total installed grid-connected solar capacity in India was 49,297 MW as of 28 February 2022, which included 42,821 MW of ground-mounted solar plants and 6,476 MW of RTS plants. Dismal growth in the RTS segment makes it imperative to identify new business models and develop new market mechanisms to drive its adoption. The regulations and framework of the Forum of Regulators emphasize the need to promote and facilitate new and innovative models for the installation of RTS systems and envisage the role of utilities in their deployment.

Playing an active role in aggregating demand, utilities can drive the adoption of RTS systems through a facilitation or investment approach. A few utility-led business models have been implemented with promising results. The guidebook discusses four business models based on which a utility can evaluate and target its RTS deployment programs according to local conditions.

Model 1: **The utility as a facilitator for the deployment of the RTS system.** It is a conventional business model being implemented by most utilities, specifically for residential consumers under Phase II of the grid-connected RTS program of the MNRE, Government of India.

Model 2: **Roof-leasing with utility investment.** The utility installs its RTS panels on roof-space leased from a consumer, who then also receives energy credits for the power generated.

Model 3: **Engineering, procurement, and construction—on an annuity payment basis, with partial stakeholder investments.** Similar to roof-leasing, this model also includes consumer contributions, thereby sharing the risks and rewards among all stakeholders—namely consumers, utilities, and developers.

Model 4: **Utility as master Renewable Energy Service Company.** This model is appropriate for high-paying consumers in long-term relationships with the utility. Such consumers are confident in allowing the utility to install and maintain the RTS system during their project's lifetime. The consumer receives power at a fixed rate, which is lower than the utility's retail supply tariff.

In August 2019, the MNRE announced Phase II of the RTS program scheme—proposing several measures to increase capacity, and specifically recognizing a larger, more active role for utilities. This guidebook will assist those utilities interested in implementing RTS projects within their license area. Describing models in detail, the guidebook presents a framework for cost–benefit analysis, and identifies the various stakeholders' roles and responsibilities.

1 INTRODUCTION

During the past decade, utility-scale, grid-connected solar photovoltaic projects have increased the penetration of renewable energy in India. By 28 February 2022, the total installed solar capacity reached 49.3 gigawatts (GW), compared to 11 megawatts (MW) in 2011.[1] Factors contributing to this rapid increase include maturing technology, market consolidation, and substantial cost reduction—as well as strong policy and regulatory support. According to a report by the Ministry of New and Renewable Energy (MNRE) on physical progress across India, by the end of February 2022, the total installed capacity of 49.3 GW included 42,821 MW of ground-mounted solar plants, and 6,476 MW of rooftop solar (RTS) plants.[2]

India has taken several initiatives for popularizing and increasing the capacity of solar generation. These include

- (i) introduction of policy and regulatory reforms at the centre and in the states;
- (ii) allocation of central financial assistance (CFA);
- (iii) active deployment of capacity through central and state tenders;
- (iv) taking promotional initiatives at the national and state level; and
- (v) bilateral and multilateral collaboration.

Institutions such as the International Solar Alliance, Solar Energy Corporation of India, and the National Institute of Solar Energy have been established for furthering the objectives of the National Solar Mission.

A. Renewable Energy Targets

By 2022, India was aiming to generate 175 GW in renewable energy, including 100 GW of solar power. Striving to achieve 40 GW from RTS projects, India has also raised its target to install 500 GW of renewable energy capacity by 2030. Figure 1 presents year-on-year tentative targets set by the MNRE in proportion to state-wise power consumption, and consequently solar power requirements, to meet the corresponding renewable purchase obligation (RPO).

Various state governments have undertaken policy and regulatory initiatives to promote RTS. This includes the provision of subsidies for specific consumer categories and net metering regulations. The adoption of RTS has been highest in the commercial and industrial (C&I) consumer category, largely driven by declining solar installation costs which have led to a decrease in solar tariffs in comparison with utility tariffs for C&I consumers. As of 28 February 2022, RTS installed capacity stood at 6,476 MW. The slow growth is attributed to several factors ranging from lack of access to finance and the disinterest of utilities to difficulty in choosing credible project developers and vendors. Furthermore, travel restrictions and limited business hours imposed by complete and partial lockdowns due to the coronavirus disease (COVID-19) pandemic have had an adverse impact on the deployment of RTS across the country. Delaying approvals of net metering, the pandemic also caused labor shortages and increased lenders' due diligence timelines. Given the low rate of deployment so far, it has become imperative to identify new business models and develop new market mechanisms to drive RTS adoption.

---

[1] Government of India, Ministry of New and Renewable Energy. Year-wise achievement of grid-connected solar power project. Available at https://mnre.gov.in/solar/solar-ongrid (accessed on 24 August 2021).

[2] Government of India, Ministry of New and Renewable Energy. Programme/Scheme wise Physical Progress in 2020-21 & Cumulative up to Aug. 2021. Available at https://mnre.gov.in/the-ministry/physical-progress (accessed on 24 August 2021).

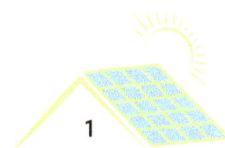

Figure 1: Year-on-Year Rooftop Solar Installation Target

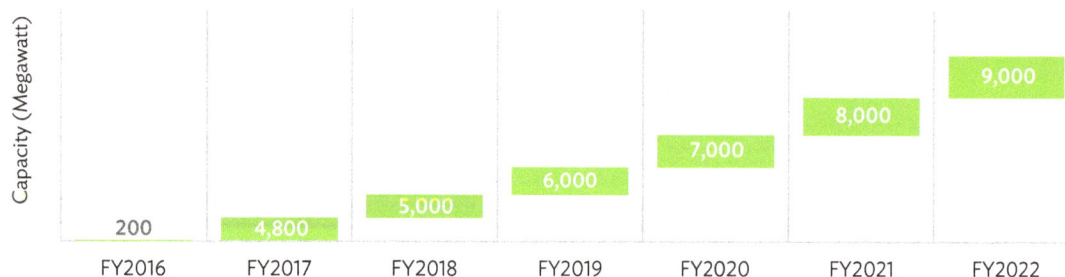

Note: "FY" before a calendar year denotes the year in which the fiscal year ends, e.g., FY2022 ends on 31 March 2022.
Source: Ministry of New and Renewable Energy, Government of India.

B. Emphasis on Utility-Led Business Models

The MNRE, Government of India, on 20 August 2019, issued the Operational Guidelines for Phase II of the Grid-Connected Rooftop Solar Program, the highlights of which are presented here.

(i) The designated implementing agency of the RTS program has been changed from the state nodal agency to the utility. It will be the responsibility of the utilities to disburse the CFA to Residential and Residential Welfare Association categories.

(ii) Incentives shall be given to the utility if the installations in a financial year are more than 10% of the base year.

(iii) The utilities shall impanel the vendors for the installation of RTS projects, and the subsidy will be disbursed to the impanel-led vendors for the plants installed.

(iv) Various business models shall be promoted under the existing legal framework such as capital expenditure (CAPEX) mode, operational expenditure (OPEX) mode, rent a roof-lease model, community model, utility model through special purpose vehicle having a share of utility, plug-in RTS model, and any other model specified by state government policy or regulations.

(v) All the states are directed to develop an online portal to process the requests of consumers applying for RTS with and without subsidy.

(vi) The utilities should also develop a grievance redress mechanism.

The RTS systems provide added benefits to the utility, viz., reduction in technical and distribution losses, reduced cost of supply to the subsidized consumer category, and better revenue streams. Involvement of the utility in the deployment enables choice in consumer category selection and dispels fear of loss of revenue from subsidizing consumer categories. At the same time, RTS systems impose several technical challenges for the utility such as voltage swell, reverse injection, and harmonics in the system. However, such challenges will be faced by the utility only after considerable penetration of the rooftop capacity in the distribution grid.

2 METERING ARRANGEMENT FOR ROOFTOP SOLAR UNDER THE PRESENT REGULATORY FRAMEWORK

Initially, only net metering and gross metering were introduced for connecting the RTS photovoltaic plant to the grid. More recently, the net billing arrangement has been added.

A. Net Metering

In case of net metering, the energy generated by the RTS plant is first allowed for self-consumption and the excess energy is injected into the grid. In case of a shortfall, the deficit energy is imported from the grid. At the end of the billing cycle, the import and export of energy are netted out at the rate fixed in the power purchase agreement (PPA) (Figure 2). At the end of the settlement cycle, net energy

- if exported to the grid, is remunerated at the average power purchase cost (APPC) or feed-in tariff (FIT) determined by the regulator; and
- if imported from the grid, is billed to the consumers at applicable utility tariffs.

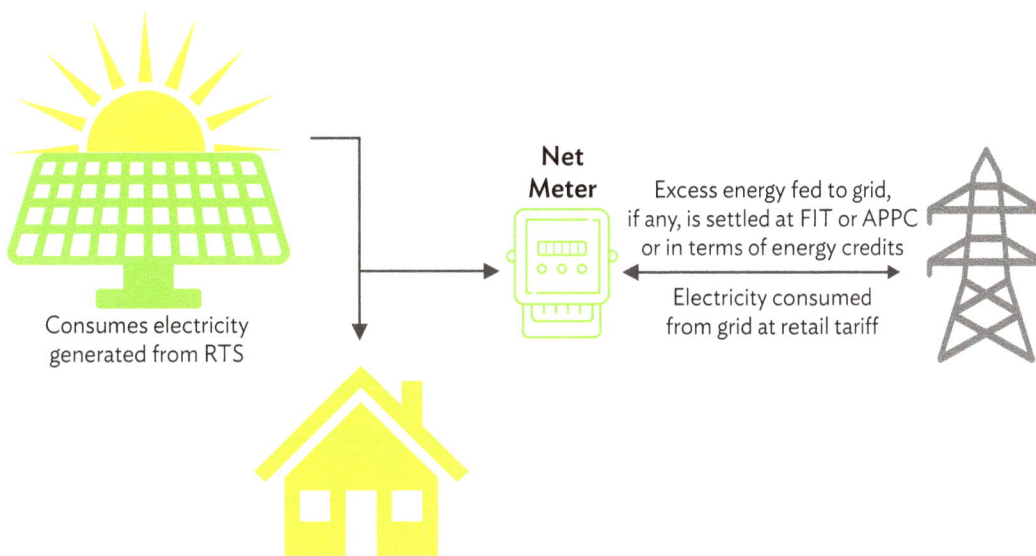

Figure 2: Schematic of Net Metering Arrangement

Consumes electricity generated from RTS

Net Meter

Excess energy fed to grid, if any, is settled at FIT or APPC or in terms of energy credits

Electricity consumed from grid at retail tariff

APPC = average power purchase cost, FIT = feed-in tariff, PPA = power purchase agreement, RTS = rooftop solar.
Source: ADB Solar Rooftop Investment Program Technical Assistance.

Under the net metering arrangement, the RTS plant is connected to the load bus of the owner. Electricity generated is primarily consumed by the owner and excess electricity, if any, is injected into the grid (Table 1).

Table 1: Illustration of Energy Accounting in Net Metering

Description		Scenario I	Scenario II
Business as usual (no solar rooftop installation)			
Total consumption in the billing period [kilowatt hour (kWh)]	A	200	100
Retail tariff (₹/kWh)	B	10	10
Consumer cash outflow (₹/month)	C = A x B	2,000	1,000
Assumption (solar rooftop system)			
Solar rooftop system capacity (kilowatt peak)	D	1	1
Number of units generated per day (kWh)	E	5	5
Settlement period (days)	F	30	30
Total generation by rooftop solar system in settlement period (kWh)	G = D x E x F	150	150
Scenario after installation of solar rooftop			
Net import from the grid during settlement period (kWh)	H If A >G, H = A- G; Else H =0	50	0
Net export to the grid during settlement period (kWh)	I If A< G, I = A – G; Else I =0	0	(-50)
Feed in tariff for injecting electricity into the grid (₹/kWh)	J	4.00	4.00
Consumer cash outflow (₹/month)	K Net import scenario: K = H x B Net export scenario: K = I x J	500	(-200)

Source: Study team analysis.

For high tension consumers, most states apply the time of day (TOD) mechanism for energy settlement. The electricity consumption in any time block (e.g., peak hours, off-peak hours, etc.) is first compensated with the electricity generation in the same time block. Any excess generation (over consumption) in any time block in a billing cycle is carried forward to the corresponding time block in the subsequent billing cycle for adjustment. In most states, the settlement period is one financial year, viz., from 1 April of the initial year to 31 March of the following year. Therefore, the excess generation is not carried forward beyond the settlement period. In many cases, no payment is made for excess generation. The power distribution companies (DISCOMs) should consider compensating excess generation at a reasonable rate after the settlement period.

B. Gross Metering

Under the gross metering arrangement, the total energy generated by the RTS plant is injected into the grid without allowing the generated solar energy to be consumed directly by the consumer. All the electricity required for consumption of the consumer is imported from the grid at applicable utility tariffs. The consumers are paid an FIT for the electricity from the RTS plant exported to the grid (Figure 3).

Figure 3: Schematic of Gross Metering Arrangement

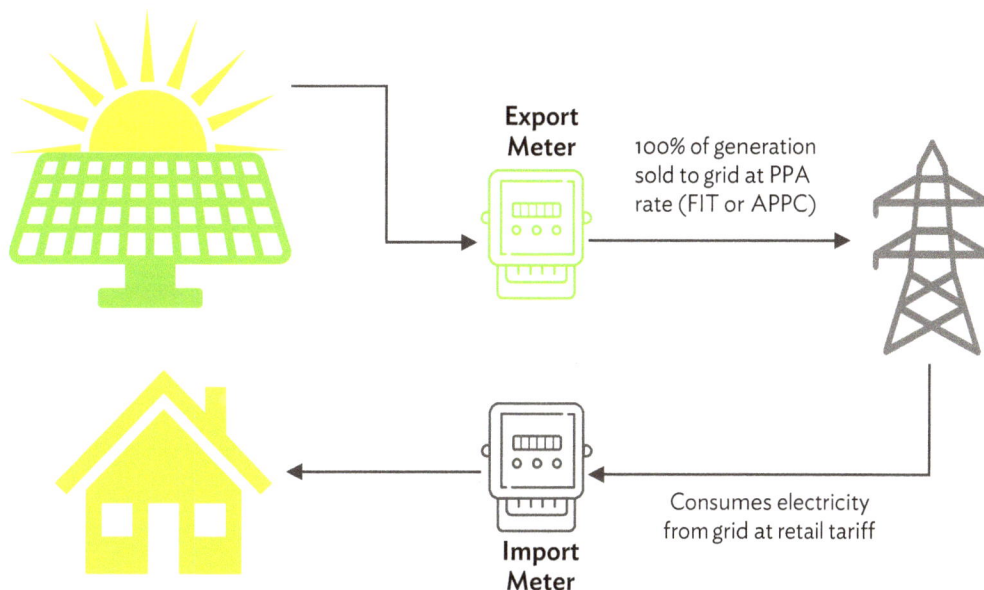

Export Meter

100% of generation sold to grid at PPA rate (FIT or APPC)

Consumes electricity from grid at retail tariff

Import Meter

APPC = average power purchase cost, FIT = feed-in tariff, , PPA = power purchase agreement.
Source: ADB Solar Rooftop Investment Program Technical Assistance.

Table 2 presents an illustration of energy accounting under the gross metering arrangement.

Table 2: Illustration of Energy Accounting in Gross Metering

Description		Scenario I	Scenario
Business as usual (no solar rooftop installation)			
Total consumption in billing period [kilowatt hour (kWh)]	A	200	100
Retail tariff (₹/kWh)	B	10	10
Consumer cash outflow (₹/month)	C = A x B	2,000	1,000
Assumption (solar rooftop system)			
Solar rooftop system capacity (kilowatt peak)	D	1	1
Number of units generated per day (kWh)	E	5	5
Settlement period (days)	F	30	30
Total generation by rooftop solar system in settlement period (kWh)	G = D x E x F	150	150
Scenario after installation of solar rooftop			
Export from rooftop solar system in settlement period (kWh)	H = G	150	150
Feed in tariff for injecting electricity in to the grid (₹/kWh)	I	4.0	4.0
Payment from distribution company for export to grid (₹/month)	J = H x I	600	600
Net consumer cash outflow (₹/month)	K = C – J	1,400	400

Source: Study team analysis.

The consumer continues to draw electricity from the grid at retail supply tariff and the generation from RTS exported to the grid is paid for by the utility to the consumer at the rate signed under the PPA, which is either FIT or APPC.

C. Net Billing

In 2019, the Forum of Regulators (FOR) introduced a new metering arrangement, net billing, under which the generation from the RTS system is fed into the grid and is measured using a unidirectional meter (y_n).[3] The power drawn by the consumer from the grid is measured through a separate unidirectional meter (x_n) (Table 3 and Figure 4). The utility then bills the consumer according to the formula,

Electricity Bill = Fixed Charges + $\Delta x \times T - \Delta y \times T'$ where,

x_n = Energy meter reading
y_n = Gross meter reading
Δx = Units consumed, i.e., $x_n - x_{n-1}$
Δy = Units generated from RTS
T = Utility tariff
T' = Net billing tariff

Table 3: Illustration of Energy Accounting in Net Billing

Description		Scenario I	Scenario II
Business as usual (no solar rooftop installation)			
Total consumption in billing period (kilowatt hour [kWh])	A	200	100
Retail tariff (₹/kWh)	B	10	10
Consumer cash outflow (₹/month)	C = A x B	2000	1000
Assumption (solar rooftop system)			
Solar rooftop system capacity (kWp)	D	1	1
Number of units generated per day (kWh)	E	5	5
Settlement period (days)	F	30	30
Total generation by rooftop solar system in settlement period (kWh)	G = D x E x F	150	150
Scenario after installation of solar rooftop			
Export from rooftop solar system in settlement period (kWh)	H = G	150	150
Net billing tariff (₹/kWh)	I	6.00	6.00
Payment from distribution company for export to grid (₹/month)	J = H x I	900	900
Net consumer cash outflow (₹/month)	K = C – J	1100	100

Source: Study team analysis.

---

3   Forum of Regulators. 2019. Metering Regulations and Accounting Framework for Grid Connected Rooftop Solar PV in India. April. Available at https://solarrooftop.gov.in/knowledge/file-32.pdf.

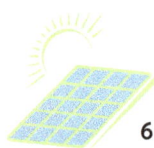

Figure 4: Schematic of Net Billing Metering Arrangement

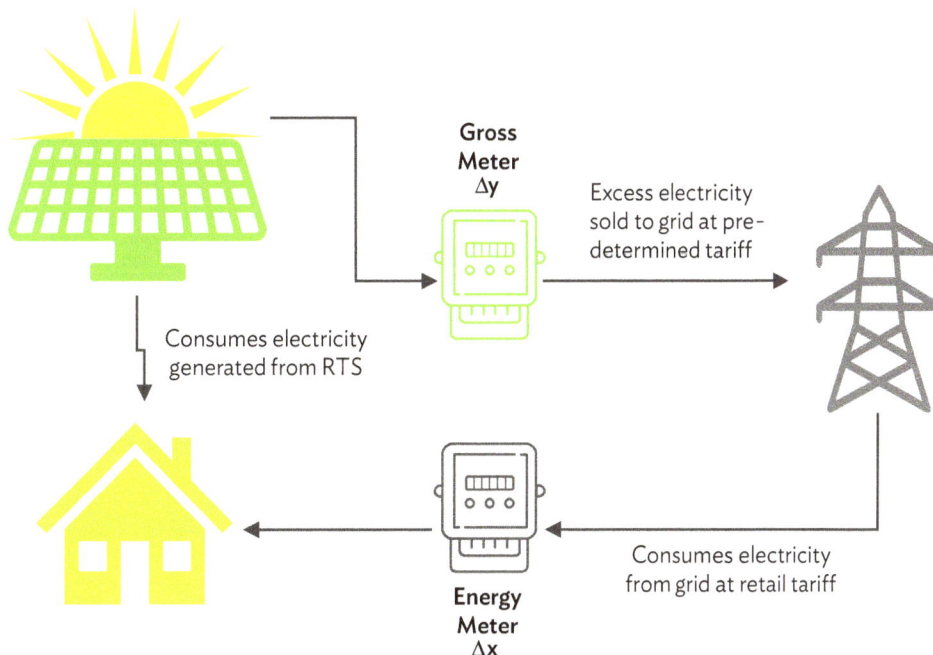

RTS = rooftop solar.
Sources:
1. Forum of Regulators. 2019. Metering Regulations and Accounting Framework for Grid Connected Rooftop Solar PV in India. April. Available at https://solarrooftop.gov.in/knowledge/file-32.pdf.
2. ADB Solar Rooftop Investment Program Technical Assistance.

The import from the grid is charged at retail supply tariff and excess generation export to the grid is paid for by the utility to the consumer at the rate signed under PPA—FIT or APPC. Further, on the metering and accounting mechanisms, the Ministry of Power, Government of India, issued the Electricity (Rights of Consumers) Rules, 2020 amendment on 28 June 2021. The salient features of the amendment (rule 11, sub-rule 4) are presented below:

- introduction of TOD tariff for projects under net metering and net billing arrangements to incentivize the consumers to install energy storage systems and feed the stored electricity into the grid during peak hours, thus enabling demand response for DISCOMs;

- installation of the solar energy meter to measure gross solar energy generated from grid interactive solar photovoltaic systems for renewable purchase obligation (RPO) credits;

- regulatory commissions to allow gross metering to all prosumers who would like to sell all the generated solar energy to the distribution licensee instead of availing of the net metering, net billing, or net feed-in facility and the commission to decide the generic tariff for gross metering as per the tariff regulations.

Status or applicability of net metering, gross metering, and net billing in different states of India as on 30 June 2022 is provided in Appendix.

3 BUSINESS MODELS FOR IMPLEMENTATION OF ROOFTOP SOLAR PROJECTS

A. Conventional Business Models

Rooftop solar systems have traditionally been set up under mainly two business models: CAPEX and OPEX (also known as Renewable Energy Service Company or RESCO).

Under the CAPEX model, the entire investment comes from the consumer. The consumer hires a solar engineering, procurement, and construction (EPC) company, which provides the turnkey installation of the entire solar power system and hands over the assets to the consumer. The company also supervises the annual operation and maintenance (O&M) of the plant on mutually agreed cost. The model could support either net or gross metering (Figures 5 and 6).

Figure 5: Capital Expenditure—Net Metering

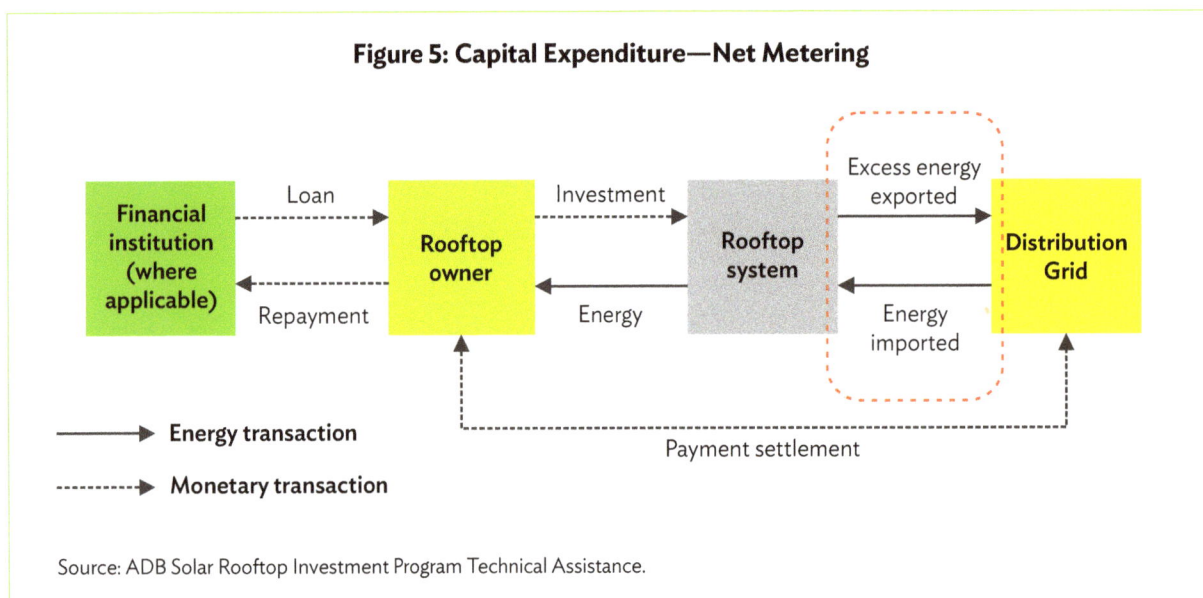

Source: ADB Solar Rooftop Investment Program Technical Assistance.

Figure 6: Capital Expenditure—Gross Metering

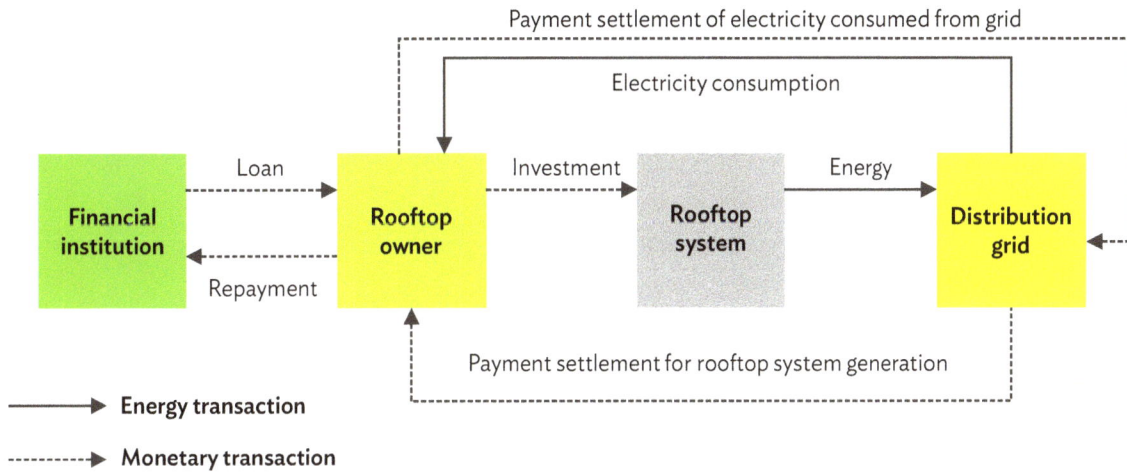

Payment settlement of electricity consumed from grid

Electricity consumption

Financial institution → Loan → Rooftop owner → Investment → Rooftop system → Energy → Distribution grid

Rooftop owner → Repayment → Financial institution

Payment settlement for rooftop system generation

→ Energy transaction
----▸ Monetary transaction

Source: ADB Solar Rooftop Investment Program Technical Assistance.

In the OPEX model, an investor or a project developer (or RESCO) invests in the CAPEX and the consumers pay for the solar power consumed from the projects developed for them. Both the consumer and the developer sign a long-term PPA for an agreed tenure and tariff. This model can also have net and gross metered variants (Figures 7 and 8).

Figure 7: Operational Expenditure—Net Metering

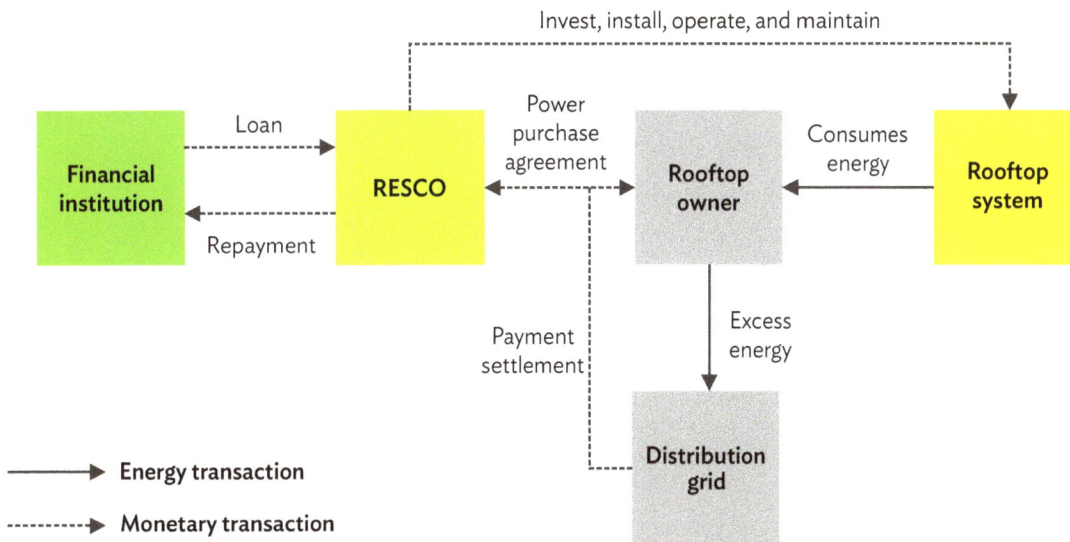

Invest, install, operate, and maintain

Power purchase agreement

Consumes energy

Financial institution → Loan → RESCO → Power purchase agreement → Rooftop owner → Consumes energy → Rooftop system

RESCO → Repayment → Financial institution

Payment settlement

Excess energy

Distribution grid

→ Energy transaction
----▸ Monetary transaction

RESCO = Renewable Energy Service Company.
Source: ADB Solar Rooftop Investment Program Technical Assistance.

Figure 8: Operational Expenditure—Gross Metering

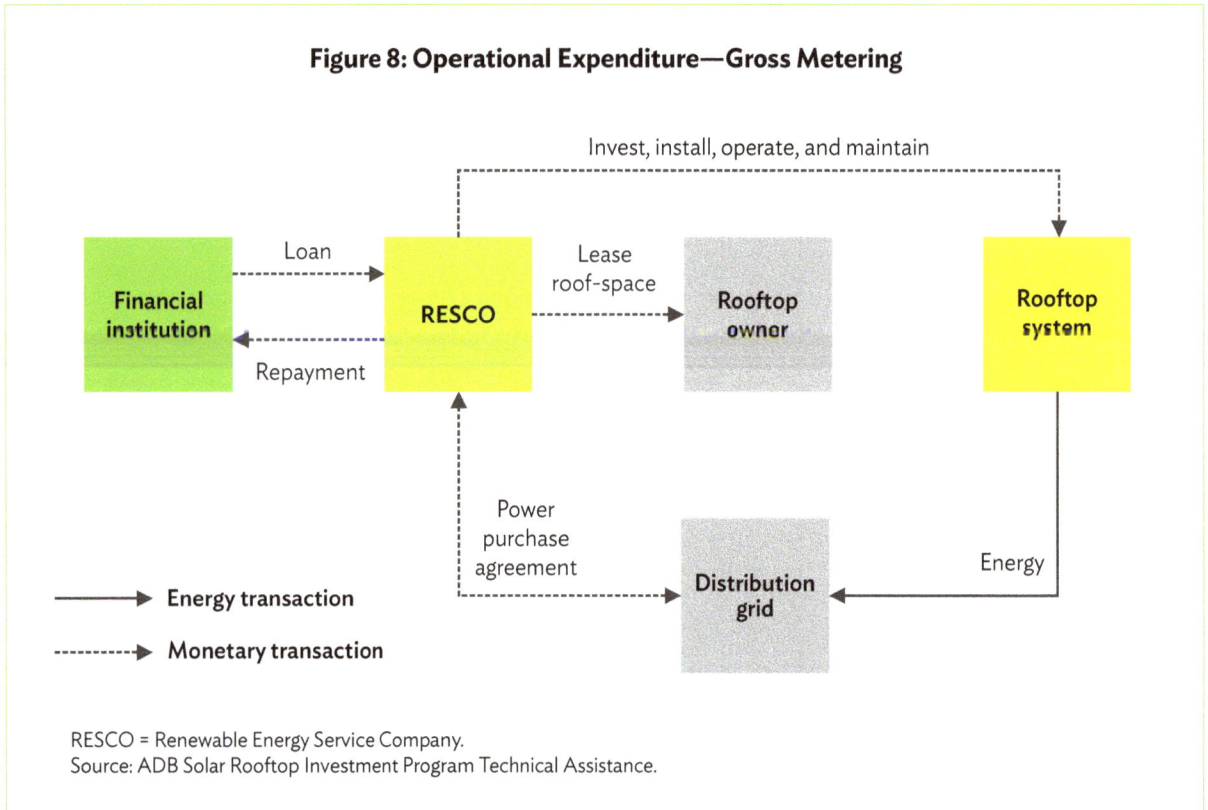

Invest, install, operate, and maintain

| Financial institution | --Loan--> | RESCO | --Lease roof-space--> | Rooftop owner | | Rooftop system |

Repayment

Power purchase agreement

Energy

Distribution grid

→ **Energy transaction**

----→ **Monetary transaction**

RESCO = Renewable Energy Service Company.
Source: ADB Solar Rooftop Investment Program Technical Assistance.

These models, however, have certain limitations which may lead to a decrease in utility revenue.

1. Utilities are generally unwilling to promote RTS in their license area as it is perceived as a threat to their incoming revenue, especially in case of net metering.

2. The residential consumers who do not come under TOD framework can use their excess solar generation exported to the grid during off-peak hours to settle the energy imported from the grid during peak time.

3. The State Electricity Regulatory Commission (SERC) sets supply tariff at rates higher than the average cost of supply of the utility to recover costs like cross subsidy, infrastructure cost, fixed charges, etc. Net metering causes loss of sales by the utility, thus, loss in revenue.

B. Utility Driven Upcoming Business Models

The CAPEX and OPEX business models under which RTS projects are currently implemented have certain limitations. For instance, it is difficult to reach out to a large number of consumers, individual rooftop sizes are small, and there are contract and payment risks, besides the fact that consumers and financial institutions lack confidence in the technology. Owing to these reasons, RTS installations have not been able to scale up.

Hence, the FOR, in the updated model regulations for grid-interactive Distributed Renewable Energy Systems has suggested promoting and facilitating new and innovative business models for scaling up RTS installations, some of which are given below:

(i) Consumer-centric CAPEX and OPEX

(ii) Utility-driven:

(a) consumer-owned (utility only aggregates);

(b) consumer-owned (utility aggregates and acts as EPC);

(c) third-party-owned (utility aggregates and acts as a trader between RESCO and consumer); or

(d) utility-owned (utility aggregates and acts as RESCO).

Utility-driven business models overcome the shortcomings of the existing business models. Due to the aggregation of demand and the scale of utility, the transaction cost is reduced, confidence is built among the consumers and financing institutions, contractual and payment risks are lowered, and the process is streamlined. Thus, it helps in scaling up the RTS installations.

4   COST–BENEFIT ASSESSMENT

Before undertaking RTS deployment, the utility needs to undertake a cost–benefit assessment of the program. The cost–benefit analysis for the utility is based on the consumer category selected and the optimum model identified for the category. In the following section, a basic framework for the estimation of cost of generation for the RTS plant is presented. This would help the utility to understand the assessment parameters and undertake the financial feasibility analysis. The cost of generation in relation to the tariff applicable to the target consumer category defines the feasibility of the model and approach selected by the utility.

A. Levelized Cost of Energy for Rooftop Solar Systems

The levelized cost of energy (LCOE) represents the average rate per unit of electricity generated and required to recover the costs of building and operating an RTS system during the project life. The LCOE is often cited as a convenient summary measure of the overall competitiveness of different generating technologies. Key inputs for the estimation of LCOE are capital costs, fuel costs, fixed and variable O&M costs, financing costs, and an assumed utilization factor for the RTS system.

1. Parameters for Calculating Levelized Cost of Energy of Rooftop Solar

a. Power Generation

(i) Capacity Utilization Factor (CUF). The CUF is the ratio of actual energy generation from the plant over the year to the maximum possible generation for a year under ideal conditions.

$$CUF = \frac{\text{Actual energy generation from solar rooftop plant in kilowatt-hour}}{(24 \times 365 \times \text{Installed capacity of rooftop solar in kilowatt})}$$

The average CUF considered by the state regulator for solar photovoltaic tariff determination ranges from 15% to 21%. Theoretically, 19% may be taken as a generic assumption. In practice though, energy generation from RTS varies from location to location.

(ii) Deration factor and module degradation factor. Deration factor is a scaling factor to account for the reduction in energy generation from the theoretical output at module level to the grid interconnection point. The deration factor considers soiling of the panels, wiring losses, temperature related losses, shading, snow cover, aging, and others. The deration factor represents system losses up to the point when the energy is fed into the grid. Moreover, the photovoltaic modules, when exposed to the atmosphere, degrade over the years and this degradation factor is defined and warranted by the module manufacturer. Most of the regulatory commissions, for generic tariff determination of RTS plants, assume linear degradation of 0.5% to the annual energy generation. However, it would be more accurate to assume 1% annual degradation of photovoltaic modules for the first 11 years and 0.71% thereon till the life of the project, viz., 25 years.

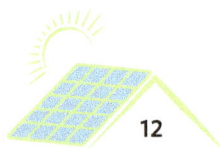

Energy generation from RTS systems is calculated as

Annual energy generation from solar rooftop plant in kilowatt-hour
= (CUF x 24 x 365 x Capacity of rooftop solar in kilowatt) – Auxilary energy consumption – deration

b.   Financial Parameters of Rooftop Solar Plant

(i)   **Capital cost.** Capital cost refers to the fixed, one time costs of designing and installing the RTS system. Capital costs are categorized into hard costs and soft costs. Hard costs are the costs of the equipment, including modules, inverters and balance of system components as well as cost of labor used in installation. Soft costs include intangible costs such as permits and taxes. Table 4 shows the capital cost considered for estimating levelized cost of energy of RTS project.

Table 4: Capital Cost for Rooftop Solar Systems (₹ per kWp)

Capacity	Capital cost
1kW	51,100
1–2 kW	46,980
2–3 kW	45,760
3–10 kW	44,640
10–100 kW	41,640
100–500 kW	39,080

kW = kilowatt, kWp = kilowatt-peak.
Source: Study team analysis.

(ii)   **Debt equity ratio.** Most state regulators follow 70:30 debt–equity ratio for the determination of generic tariff for RTS systems.

(iii)   **Moratorium period.** Moratorium period of 6–12 months is offered by banks.

(iv)   **Repayment period.** Repayment period of up to 13 years is allowed for financing of RTS projects.

(v)   **Interest rate on debt.** Lenders offer 1-year marginal cost of funds based lending rate plus spread in the range of 50–200 basis points based on the risk rating of the consumer.

(vi)   **Return on equity (ROE).** Central Electricity Regulatory Commission defined the normative ROE as 14%,[4] to be grossed up by prevailing minimum alternate tax (MAT) as on 1 April of the previous year for the entire useful life of the project (Figure 9).

Therefore, posttax ROE is assumed to be 14%. Pretax ROE is calculated based on corporate tax rates and minimum alternate tax rates. Table 5 provides the breakup of taxes applicable to RTS projects.

---

[4]   http://www.cercind.gov.in/2017/orders/05.pdf.

Figure 9: Calculation of Return on Equity

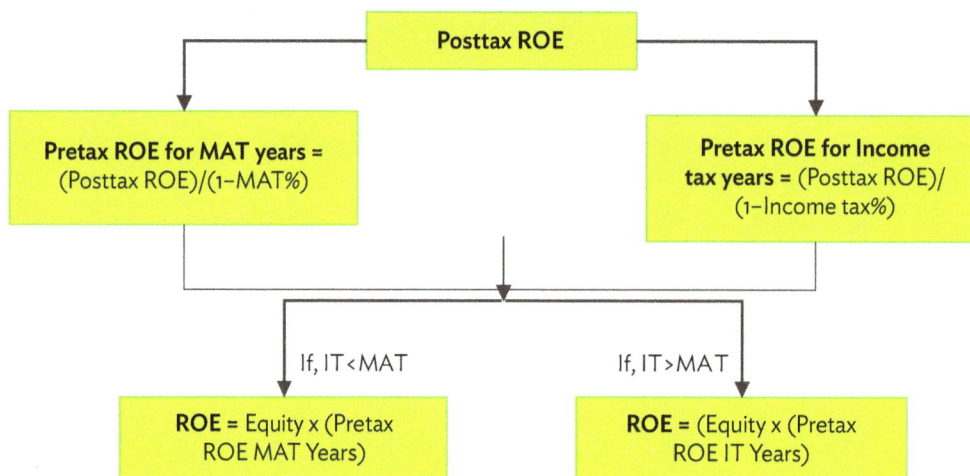

```
                            ┌─────────────────────┐
                            │     Posttax ROE     │
                            └─────────────────────┘
        ┌───────────────────────────┐      ┌───────────────────────────┐
        │  Pretax ROE for MAT years =│      │  Pretax ROE for Income    │
        │   (Posttax ROE)/(1–MAT%)   │      │  tax years = (Posttax ROE)/│
        │                            │      │      (1–Income tax%)       │
        └───────────────────────────┘      └───────────────────────────┘

                    If, IT<MAT                      If, IT>MAT

        ┌───────────────────────────┐      ┌───────────────────────────┐
        │  ROE = Equity x (Pretax    │      │  ROE = (Equity x (Pretax   │
        │       ROE MAT Years)       │      │        ROE IT Years)       │
        └───────────────────────────┘      └───────────────────────────┘
```

IT = income tax, MAT = minimum alternate tax, ROE = return on equity.
Source: Government of India.

Table 5: Income Tax and Minimum Tax

Income tax	Percentage
Corporate tax	22.00
Surcharge	10.00
Base rate + surcharge	24.20
Health and educational cess	4.00
Base rate+ surcharge + education cess	25.17
Total tax rate	25.17
Minimum Alternate Tax (MAT)	
Period for MAT credit (years)	10
Base rate	15.00
Surcharge	10.00
Base rate + surcharge	16.50
Health and educational cess	4.00
Base rate + surcharge + education cess	17.16
Total tax rate	17.16

Source: Income tax department, Government of India.

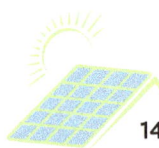

Table 6: Summary of the Parameters Used for Estimating Levelized Cost of Energy

Parameter	Assumptions
Capital cost for a 100–500 kilowatt peak (kWp) RTS system	₹39,080 per kWp
Capacity utilization factor	19%
Auxiliary consumption	0%
Deration factor	1% for first 11 years; 0.71% from 12th year onward
Tariff Period and Plant Life	25 years
Debt–equity ratio	70:30
Moratorium period	6 months
Repayment period	13 years
Interest rate	9% (MCLR + 40 basis points)
Posttax ROE	14.00%
Pretax ROE–MAT years	17.85%
Pretax ROE–income tax years	19.75%
Discount rate (weighted average cost of capital)	8.91%
Income tax	25.17%
MAT rate	17.16%
Book depreciation	5.28% for first 13 years; 1.78% after 13 years
Maximum book depreciation	90%
O&M cost	₹600,000 per MW
O&M cost escalation	5.72%
Interest on working capital	10% (MCLR + 140 basis points)

MAT = minimum alternate tax, MCLR = marginal cost of funds based lending rate, MW = megawatt, O&M = operation and maintenance, ROE = return on equity.

Source: Order on the generic tariff determination for rooftop solar of the Government of India, Ministry for New and Renewable Energy or State Electricity Regulatory Commission (SERC) as applicable.

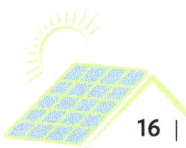

Table 7: Determination of Levelized Cost of Energy for Rooftop Solar Plants Aggregating to 1 Megawatt

Units generated	Year --> 1	2	3	4	5	6	7	8	9	10	11	12	13	14	15	16	17	18	19	20	21	22	23	24	25
Installed capacity (megawatt)	1	1	1	1	1	1	1	1	1	1	1	1	1	1	1	1	1	1	1	1	1	1	1	1	1
Capacity utilization factor (%)	19	19	19	19	19	19	19	19	19	19	19	19	19	19	19	19	19	19	19	19	19	19	19	19	19
Gross generation (mega unit or MU)	1.66	1.66	1.66	1.66	1.66	1.66	1.66	1.66	1.66	1.66	1.66	1.66	1.66	1.66	1.66	1.66	1.66	1.66	1.66	1.66	1.66	1.66	1.66	1.66	1.66
Auxiliary consumption (MU)	0	0	0	0	0	0	0	0	0	0	0	0	0	0	0	0	0	0	0	0	0	0	0	0	0
Deration loss (MU)	–	0.02	0.03	0.05	0.07	0.08	0.10	0.12	0.13	0.15	0.17	0.18	0.19	0.20	0.21	0.23	0.24	0.25	0.26	0.27	0.29	0.30	0.31	0.32	0.33
Net generation (MU)	1.66	1.65	1.63	1.61	1.60	1.58	1.56	1.55	1.53	1.51	1.50	1.49	1.47	1.46	1.45	1.44	1.43	1.41	1.40	1.39	1.38	1.37	1.36	1.34	1.33
Fixed cost (₹ million)	Year --> 1	2	3	4	5	6	7	8	9	10	11	12	13	14	15	16	17	18	19	20	21	22	23	24	25
Operation and maintenance expenses	–	–	–	–	–	0.60	0.63	0.67	0.71	0.75	0.79	0.84	0.89	0.94	0.99	1.05	1.11	1.17	1.24	1.31	1.38	1.46	1.54	1.63	1.73
Return on equity	1.98	1.98	1.98	1.98	1.98	1.98	1.98	2.19	2.19	2.19	2.19	2.19	2.19	2.19	2.19	2.19	2.19	2.19	2.19	2.19	2.19	2.19	2.19	2.19	2.19
Interest on debt	2.37	2.18	1.99	1.80	1.61	1.42	1.23	1.04	0.85	0.66	0.47	0.28	0.09	0.00	0.00	0.00	0.00	0.00	0.00	0.00	0.00	0.00	0.00	0.00	0.00
Loan repayment	2.10	2.10	2.10	2.10	2.10	2.10	2.10	2.10	2.10	2.10	2.10	2.10	2.10	–	–	–	–	–	–	–	–	–	–	–	–
Interest on working capital	0.10	0.10	0.10	0.09	0.09	0.11	0.11	0.11	0.11	0.11	0.11	0.11	0.11	0.11	0.11	0.11	0.11	0.11	0.11	0.11	0.11	0.11	0.12	0.12	0.12
Total fixed cost	6.55	6.36	6.17	5.98	5.79	6.21	6.06	6.12	5.96	5.82	5.67	5.53	5.38	3.24	3.29	3.35	3.41	3.47	3.54	3.61	3.69	3.77	3.85	3.94	4.04

continued on next page

Table 7: continued

Levelized cost of generation (₹ per kilowatt-hour)	Year --> 1	2	3	4	5	6	7	8	9	10	11	12	13	14	15	16	17	18	19	20	21	22	23	24	25
Operation and maintenance expenses	–	–	–	–	–	0.38	0.41	0.43	0.46	0.49	0.53	0.56	0.60	0.64	0.68	0.73	0.78	0.83	0.88	0.94	1.00	1.07	1.14	1.22	1.30
Return on equity	1.19	1.20	1.21	1.23	1.24	1.25	1.27	1.42	1.43	1.45	1.46	1.48	1.49	1.50	1.51	1.52	1.54	1.55	1.56	1.58	1.59	1.60	1.62	1.63	1.65
Interest on debt	1.42	1.32	1.22	1.11	1.01	0.90	0.79	0.67	0.56	0.44	0.32	0.19	0.06	0.00	0.00	0.00	0.00	0.00	0.00	0.00	0.00	0.00	0.00	0.00	0.00
Loan repayment	1.26	1.28	1.29	1.30	1.32	1.33	1.35	1.36	1.37	1.39	1.40	1.42	1.43	–	–	–	–	–	–	–	–	–	–	–	–
Interest on working capital	0.06	0.06	0.06	0.06	0.06	0.07	0.07	0.07	0.07	0.07	0.07	0.07	0.07	0.07	0.07	0.08	0.08	0.08	0.08	0.08	0.08	0.08	0.09	0.09	0.09
Total cost of generation	3.94	3.86	3.78	3.70	3.62	3.93	3.87	3.95	3.90	3.84	3.78	3.72	3.65	2.21	2.27	2.33	2.39	2.45	2.52	2.60	2.67	2.76	2.84	2.94	3.03
Discount factor	1.00	0.92	0.84	0.77	0.71	0.65	0.60	0.55	0.51	0.46	0.43	0.39	0.36	0.33	0.30	0.28	0.26	0.23	0.22	0.20	0.18	0.17	0.15	0.14	0.13
Levelized tariff	3.51																								

Source: ADB Solar Rooftop Investment Program Technical Assistance

B. Estimation of Benefits to the Utility

Utility benefits are drawn from savings in power purchase costs and renewable purchase obligation (RPO) from the RTS system. Utility loses revenue in case the rooftop system is owned by the consumer.

Benefit to DISCOMs = Annual generation from RTS plant x ([Average power purchase cost with annual escalation inclusive of distribution losses] + [Renewable energy certificate floor price] – [Energy charges applicable to consumer category with annual escalation])

This is explained through an example of the Kerala utility. Table 8 highlights assumptions made for estimating the benefit to Kerala State Electricity Board Limited (KSEDL). Benefit to KSEDL is illustrated in Table 11.

Table 8: Assumptions for Estimating Benefit to Kerala State Electricity Board Limited

Parameter	Unit	Value
Consumer category		Domestic
Average revenue realized	₹ per kWh	4.62
Escalation in utility tariff	%	2.00
Variable charge rate of marginal plants	₹ per kWh	4.87
Escalation of variable charges of marginal power plants	%	2.00
Renewable energy certificate floor price	₹ per kWh	1

kWh = kilowatt-hour.
Sources:
1. Kerala State Electricity Board. 2019. Aggregate Revenue Requirement Order of the Kerala State Electricity Board for FY2022. https://www.kseb.in/index.php?option=com_jdownloads&task=download.send&id=10968&catid=64&m=0&lang=en.
2. ADB Solar Rooftop Investment Program Technical Assistance.

Additionally, distribution companies (DISCOMs) are eligible for an incentive for promoting and enabling the deployment of RTS under the MNRE Phase II grid-connected RTS scheme. Incentives are calculated on the basis of installed base capacity (Table 9).

Table 9: Incentives Available to Distribution Companies under the Rooftop Solar Scheme of the Ministry of New and Renewable Energy, Government of India

	Parameter	Incentives to be provided
1	For installed capacity achieved 10%–15% over and above the installed base capacity[a] within a financial year	5% of the applicable cost[b] for capacity achieved above 10% of the installed base capacity

continued on next page

Table 9: continued

	Parameter	Incentives to be provided
2	**For installed capacity achieved beyond 15% over and above the installed base capacity[a] within 1 financial year**	5% of the applicable cost[b] for capacity achieved 10%–15% over and above the installed base capacity plus 10% of the applicable cost[b] for capacity achieved beyond 15% of the installed base capacity

DISCOM = Distribution company, kW = kilowatt, MNRE = Ministry of New and Renewable Energy, PSU = Public sector undertaking, RTS = Rooftop Solar.
[a] Installed base capacity shall mean the cumulative RTS capacity installed within the jurisdiction of DISCOMs at the end of previous financial year. This will include total RTS capacity installed under residential, institutional, social, government, PSU, statutory/ autonomous bodies, private commercial, industrial sectors etc.
[b] Applicable cost is the applicable benchmark cost of MNRE for the states/Union Territories for mid-range RTS capacity of above 10 kW and up to 100 kW or lowest of the costs discovered in the tenders for that states/Union Territories in that year, whichever is lower.
Source: Government of India, Ministry of New and Renewable Energy. 2019. Guidelines on implementation of Phase – II of Grid Connected Rooftop Solar Programme for achieving 40 GW capacity from Rooftop Solar by the year 2022. https://mnre.gov.in/img/documents/uploads/7ccd3b4b3bb94a51af516e2ee4fdede3.pdf.

Table 10 illustrates the incentive mechanism for DISCOMs. Table 11 presents the calculation of benefit to the utility from the installation of a 1MW RTS where the consumer invests.

Table 10: Illustration of Incentive Mechanism

Baseline capacity or installed capacity in previous fiscal year (MW)	Capacity installed in current fiscal year (MW)	Percentage achievement of installed base capacity (%)	Capacity eligible for 5% incentives (MW)	Capacity eligible for 10% incentives (MW)
100	10	10	Nil	Nil
100	15	15	5	Nil
100	30	30	5	15
100	50	50	5	35

MW = megawatt.
Source: ADB Solar Rooftop Investment Program Technical Assistance.

Table 11: Benefit to Utility upon Installation of 1 Megawatt Rooftop Solar Plant

Year-->	1	2	3	4	5	6	7	8	9	10	11	12
Grid-connected rooftop solar plant capacity (megawatt)	1.00	1.00	1.00	1.00	1.00	1.00	1.00	1.00	1.00	1.00	1.00	1.00
Capacity utilization factor (%)	18	18	18	18	18	18	18	18	18	18	18	18
Energy injected to grid (million unit)	1.58	1.58	1.58	1.58	1.58	1.58	1.58	1.58	1.58	1.58	1.58	1.58
Utility energy charges (₹ per kilowatt-hour)	4.62	4.71	4.81	4.90	5.00	5.10	5.20	5.31	5.41	5.52	5.63	5.74
Loss in utility revenue (₹ million)	-7.28	-7.43	-7.58	-7.73	-7.89	-8.04	-8.20	-8.37	-8.54	-8.71	-8.88	-9.06
Variable charge rate of marginal power plants (₹ per kilowatt-hour)	4.87	4.97	5.07	5.17	5.27	5.38	5.48	5.59	5.71	5.82	5.94	6.06
Savings due to avoided marginal power plants (₹ million)	7.68	7.83	7.99	8.15	8.31	8.48	8.65	8.82	9.00	9.18	9.36	9.55
Avoided cost of purchasing renewable energy certificates (₹ million)	1.58	1.58	1.58	1.58	1.58	1.58	1.58	1.58	1.58	1.58	1.58	1.58
Gross savings (₹ million)	1.97	1.98	1.99	2.00	2.00	2.01	2.02	2.03	2.04	2.05	2.06	2.07
Net present value of gross savings (₹ million)	20.62	–	–	–	–	–	–	–	–	–	–	–

continued on next page

Table 11: continued

Year-->	13	14	15	16	17	18	19	20	21	22	23	24	25
Grid-connected rooftop solar plant capacity (megawatt)	1.00	1.00	1.00	1.00	1.00	1.00	1.00	1.00	1.00	1.00	1.00	1.00	1.00
Capacity utilization factor (%)	18	18	18	18	18	18	18	18	18	18	18	18	18
Energy injected to grid (million unit)	1.58	1.58	1.58	1.58	1.58	1.58	1.58	1.58	1.58	1.58	1.58	1.58	1.58
Utility energy charges (₹ per kilowatt-hour)	5.86	5.98	6.10	6.22	6.34	6.47	6.60	6.73	6.87	7.00	7.14	7.29	7.43
Loss in utility revenue (₹ million)	-9.24	-9.42	-9.61	-9.80	-10.00	-10.20	-10.40	-10.61	-10.82	-11.04	-11.26	-11.49	-11.72
Variable charge rate of marginal power plants (₹ per kilowatt-hour)	6.18	6.30	6.43	6.55	6.69	6.82	6.96	7.09	7.24	7.38	7.53	7.68	7.83
Savings due to avoided marginal power plants (₹ million)	9.74	9.93	10.13	10.33	10.54	10.75	10.97	11.19	11.41	11.64	11.87	12.11	12.35
Avoided cost of purchasing renewable energy certificates (₹ million)	1.58	1.58	1.58	1.58	1.58	1.58	1.58	1.58	1.58	1.58	1.58	1.58	1.58
Gross savings (₹ million)	2.08	2.09	2.10	2.11	2.12	2.13	2.14	2.15	2.16	2.17	2.19	2.20	2.21
Net present value of gross savings (₹ million)	–	–	–	–	–	–	–	–	–	–	–	–	–

Note: Incentives under the Phase II grid-connected Rooftop Solar scheme of the Ministry of New and Renewable Energy are not included as they are dependent on the baseline capacity of distribution companies.
Source: ADB Solar Rooftop Investment Program Technical Assistance.

5 UTILITY-LED BUSINESS MODELS FOR ROOFTOP SOLAR DEPLOYMENT

Rooftop solar (RTS) systems offer several advantages to the utility.

(i)   They do not require additional land for electricity generation.

(ii) They offer reduced transmission and distribution losses due to closer proximity of loads from the generation units.

(iii) The utility can tailor the business model to the needs of the targeted consumer category.

(iv) The utility can reduce its cost of service to subsidized consumer categories.

Thus, it makes economic sense for the utilities to aggregate the demand for RTS and take an active role in its deployment. Phase II of the MNRE RTS program also incentivizes utilities to achieve additional grid-connected RTS capacity. Various business models which can be adopted by the utilities are described in the following sections.

Conventional business models depend on consumers to invest in and deploy RTS under net metering, gross metering, or net billing arrangement, as defined by state regulations. While RTS proved beneficial to consumers (especially the high-paying ones), utilities perceived a direct risk to their revenue from RTS deployment. Under the Phase II of MNRE RTS program, however, since the utilities are the implementing agencies, they can play a more active role by investing in RTS or implementing business models targeting specific consumer categories based on the benefit to utilities as well as consumers.

The following business models are proposed for the deployment of RTS with the utilities as primary stakeholders:

Model 1: Utility as Facilitator for Deployment of RTS

Model 2: Roof-Leasing with Utility Investment

Model 3: EPC–Annuity Payment Basis with Partial Stakeholder Investments

Model 4: Utility as Master RESCO

A. Model 1: Utility as Facilitator for the Deployment of Rooftop Solar

1. Model Outline

Under the facilitation approach, the consumer invests and installs the RTS plant, whereas the utility undertakes the price discovery along with the standardization of technical specifications for the system.

The utility may consider two options under this model.

(i) The utility invites interest from consumers for RTS deployment and after identification of buildings and undertaking feasibility assessment, bids out the capacity to rooftop EPC companies. For building identification and feasibility assessment, utilities can charge a facilitation fee to the EPC companies.

(ii) The utility can impanel EPC companies and discover price for installation of RTS systems. Consumers select an EPC from the list.

The first option offers the advantage of selecting a specific consumer category to the utility. The consumers' payments to the EPC companies can be routed through the utility and central financial assistance can be handled and disbursed by the utility, thereby streamlining the complete payment system for both the consumers as well as the EPC companies.

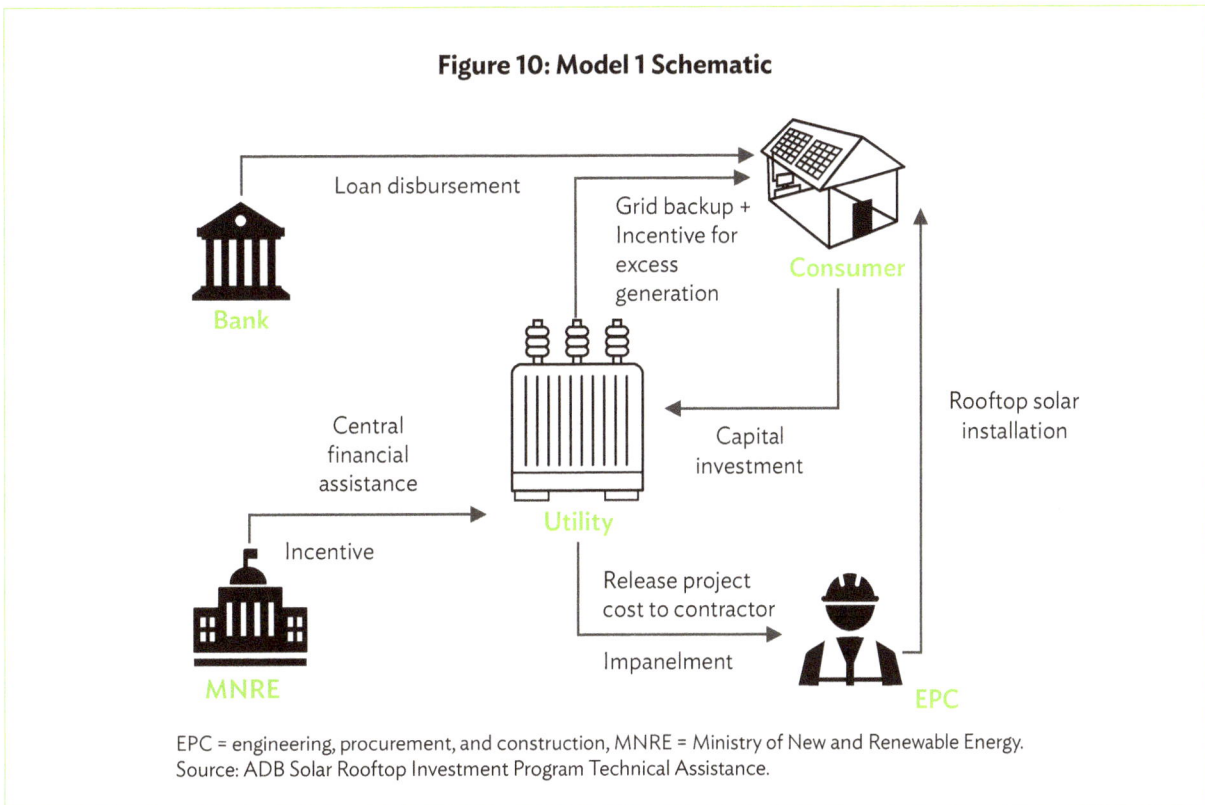

Figure 10: Model 1 Schematic

EPC = engineering, procurement, and construction, MNRE = Ministry of New and Renewable Energy.
Source: ADB Solar Rooftop Investment Program Technical Assistance.

2. Roles and Responsibilities of Stakeholders

Table 12 depicts the roles and responsibilities of various stakeholders under the described model.

Table 12: Roles and Responsibilities of Stakeholders

Consumer	Utility	Developer
(i) Invest in rooftop solar system.	(i) Aggregate interest from the consumer.	(i) Undertake design, supply, installation, metering, testing, and commissioning.
(ii) Enter into interconnection agreement with the utility.	(ii) Avail and facilitate subsidy from the Ministry of New and Renewable Energy, Government of India and the state government, if applicable.	(ii) Perform installations and allied work up to the interconnection point.
(iii) Ensure safety and security of the system.	(iii) Select EPC companies through competitive bidding.	(iii) Provide O&M for 5 years from the date of project commissioning.
(iv) Undertake regular O&M post contract period with EPC.	(iv) Facilitate transaction between the consumer and EPC.	
	(v) Monitor the installation of rooftop solar on consumer's roof.	

EPC = engineering, procurement, and construction; O&M = operation and maintenance.
Source: ADB Solar Rooftop Investment Program Technical Assistance.

3. Financial Impact on Stakeholders

Figure 11 illustrates the cost–benefit analysis for the consumers and utility.

Figure 11: Cost–Benefit Analysis for Consumers and Utility

	Consumer	Utility
COSTS	• Capital cost of installing GRPV • Facilitation fee to utility • O&M after contract with developers	• Cost of tendering and facilitation • Loss of revenue due to reduction in sale of power to consumers
BENEFITS	• Savings on utility bill based on the installed GRPV capacity	• Reduction in T&D losses • Savings in power purchase cost from marginal power plants • RPO benefit, if applicable • Incentives and service charges under MNRE Phase-II solar rooftop scheme

GRPV = grid-connected rooftop solar photovoltaic, O&M = operation and maintenance,
MNRE = Ministry of New and Renewable Energy, RPO = renewable purchase obligation, T&D = transmission and distribution.
Source: ADB Solar Rooftop Investment Program Technical Assistance.

Considering the gap in power purchase from the marginal power plant and the average billing rate (ABR) of low paying consumers, adoption of the RTS by low-paying consumers helps utilities save on loss in average cost of supply, in turn reducing cross subsidy on high-paying consumers. As the cost of generation from the RTS is much lower than the energy charges paid by high-paying consumers on their utility bill, adoption of the RTS by high-paying consumers impacts the revenue of utility adversely.

High-paying consumers are the primary revenue generators for utilities and low-paying consumers are cross subsidized. Therefore, it is important for the utility to serve low-paying consumers along with continuing service to high-paying consumers to maintain the balance in revenue requirement of the utility.

4. Assumptions for Cost–Benefit Assessment

The following parameters are to be assessed to understand the benefit to utilities:

(i)   loss of revenue to utility (ABR or revenue realization);

(ii) benefit of savings in power purchase considered at average power purchase cost (APPC);

(iii) savings in distribution losses due to localized consumption of power; and

(iv) avoidance of renewable energy certificates (REC) purchase by utilities which are not meeting RPO targets.

To assess the benefits to the utility, the case of Paschim Gujarat Vij Company Ltd, Gujarat is presented. State-specific assumptions considered are shown in Table 13. The assumptions for power generation from RTS have already been presented in Table 6.

Table 13: Assumptions for Cost–Benefit Analysis

Parameter[a]	Value
Average energy charges revenue for domestic consumers	₹3.65 per kWh
Average energy charges revenue for commercial consumers[b]	₹5.45 per kWh
Average energy charges revenue for industrial consumers	₹5.50 per kWh
Escalation in energy charges	2%
Average power purchase cost (APPC)[c]	₹4.10 per kWh
Distribution losses	16%
APPC inclusive of distribution losses	₹4.76 per kWh
Escalation in APPC	2%
Renewable energy certificate floor price	₹1.00 per kWh

kWh = kilowatt-hour.
[a] Parameters are assumed based on the Aggregate Revenue Requirement order of Paschim Gujarat Vij Company Limited (PGVCL) for FY2022.
[b] Tariff applicable to nonresidential general purpose and low tension maximum demand consumer category is considered under commercial consumers.
[c] Gujarat Urja Vikas Nigam Ltd is the central power procurement agency for all four DISCOMs of Gujarat. Average power purchase cost of PGVCL is calculated based on the amount available for power purchase or energy purchased.
Note: The cost of generation of the RTS plant is calculated based on the assumptions considered in Tables 4 and 6.
Sources:
1. Gujarat Electricty Regulatory Commission. 2021. Aggregate Revenue Requirement order of Paschim Gujarat Vij Company Limited (PGVCL) for FY2022. https://gercin.org/wp-content/uploads/2021/04/PGVCL-1914-2020-True-up-FY-2019-20-Order-31.03.2021.pdf.
2. ADB Solar Rooftop Investment Program Technical Assistance.

5. Benefit to Utility

Based on the above assumptions, impact on the utility is assessed by the following formula:

> Impact on Utility = Total generation x ([APPC inclusive of distribution losses with escalation] + [Savings from purchase of RECs, if applicable] – [Energy charges for specific consumer category with escalation]).

Figure 12 shows the impact on the utility under this business model for the installation of an RTS of 1MW aggregated capacity for a specific consumer category.

(i) It is observed that utility gains are considerable upon the deployment in the residential consumer category.

(ii) The utility faces revenue loss in high-paying consumers.

(iii) The utility will gain a one-time incentive from the MNRE and facilitation fee from the developer.

Figure 12: Impact of Rooftop Solar Installation under Various Consumer Categories on Distribution Companies

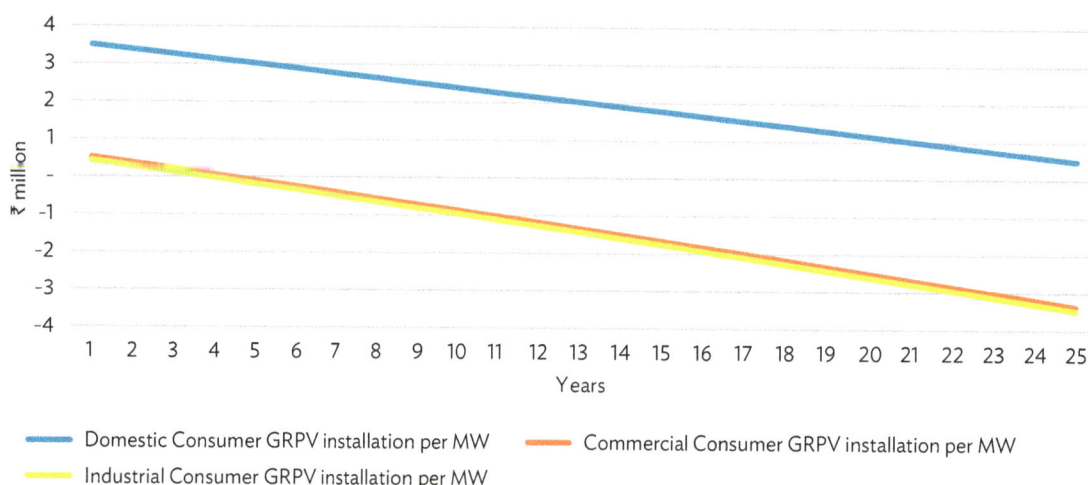

GRPV = grid-connected rooftop solar photovoltaic, MW = megawatt.
Sources:
1. ADB Solar Rooftop Investment Program Technical Assistance.
2. Aggregate Revenue Requirement order of Paschim Gujarat Vij Company Limited for FY2022.
   Available at https://gercin.org/wp-content/uploads/2022/04/PGVCL-20312021-Tariff-Order-of-FY-2022-23-dtd.-31.03.2022.pdf.

6. Benefit to Consumers

An analysis of consumer benefit, assuming the RTS system is installed under net metering arrangement and power generation is used by the consumer for self-consumption, is shown in Table 14. To evaluate the benefit to consumers, for 1–10 kilowatt-peak (kWp) RTS plants, utility tariff of the domestic consumer (₹3.65 per kilowatt-hour [kWh]) is considered. For RTS plants above 10kWp, industrial tariff (₹5.50 per kWh) is considered with an annual escalation of 2%.

Table 14: Benefit to Consumer

Parameters	Unit	Return on investment			
		Rooftop solar capacity			
		1kWp	>3–10kWp	>10–100kWp	>100–500kWp
Capital cost[a]	₹ per kWp	51,100	44,640	41,640	39,080
Subsidy	%	40%	26%–35%	–	–
Investment by consumer	₹ per kWp	30,660	31,248[b]	41,640	39,080
Cost of generation	₹ per kWh	2.82	2.87	3.71	3.51
Payback period	Years	3.20	3.26	3.07	2.32

kWh = kilowatt-hour, kWp = kilowatt-peak.

[a] The capital cost is exclusive of net meter cost.

[b] Average central financial assistance of 30% is assumed for rooftop solar capacity 3–10 kWp.

Source: ADB Solar Rooftop Investment Program Technical Assistance.

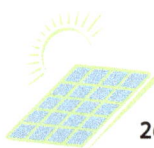

7. Key Features of Model 1

(i) The utility aggregates the RTS demand and undertakes price discovery but the investment is made by the consumer.

(ii) The utility facilitates subsidy disbursement for the consumers.

(iii) The utility can adopt this model for any consumer category.

Box 1: Outreach Activities under SURYA Gujarat

Gujarat is the leading state in rooftop solar (RTS) deployment. It has achieved 36% (1,140 megawatts [MW] as of May 2021) of its 3,600 MW target, of which 800 MW has been deployed in the past 2 years. The achievement was made possible by the multiple measures adopted by distribution companies (DISCOMs) of Gujarat for the promotion of RTS and awareness of domestic consumers. These measures include

- implementation of RTS portal and digitizing the interconnection and subsidy process for bringing transparency and improving the ease of doing business;

- communicating with consumers through SMS and missed call initiatives;

- disseminating key information on RTS systems;

- dedicated and timely follow-up by designated DISCOM officials on a regular basis;

- broadcasting audio advertisements in government and private radio stations during peak listening hours;

- display of flexi banners across all government offices in Gujarat;

- information flyers distributed to residential consumers along with electricity bills;

- public awareness programs arranged in town halls, municipalities, and municipal corporations;

- publishing full page multilingual advertisements;

- public hoardings displayed in prominent places of regular public visits;

- advertisements on TV channels and cinema theaters; and

- social media engagement.

Source: Inputs received from Gujarat Urja Vikas Nigam Limited and Deutsche Gesellschaft für Internationale Zusammenarbeit (GIZ) GmbH

B. Model 2: Roof-Leasing with Utility Investment

1. Model Outline

Under this model, the utility invests in, installs, commissions, and maintains the RTS plant on the consumer's roof under gross-metering arrangement. The consumer receives a rebate on the utility bill in terms of energy credits equivalent to a part of the RTS generation. Further, a minimum lock-in period of 5 years is imposed on the consumer to ensure safety of the investment made by utility. The rebate to the consumer is subject to the cost–benefit assessment for all stakeholders. The consumer is expected to ensure safety and security of RTS plant. Figure 13 shows the transactions across all stakeholders under this model.

Figure 13: Model 2 Schematic

EPC = engineering, procurement, and construction; MNRE = Ministry of New and Renewable Energy;
O&M = operation and maintenance.
Source: ADB Solar Rooftop Investment Program Technical Assistance.

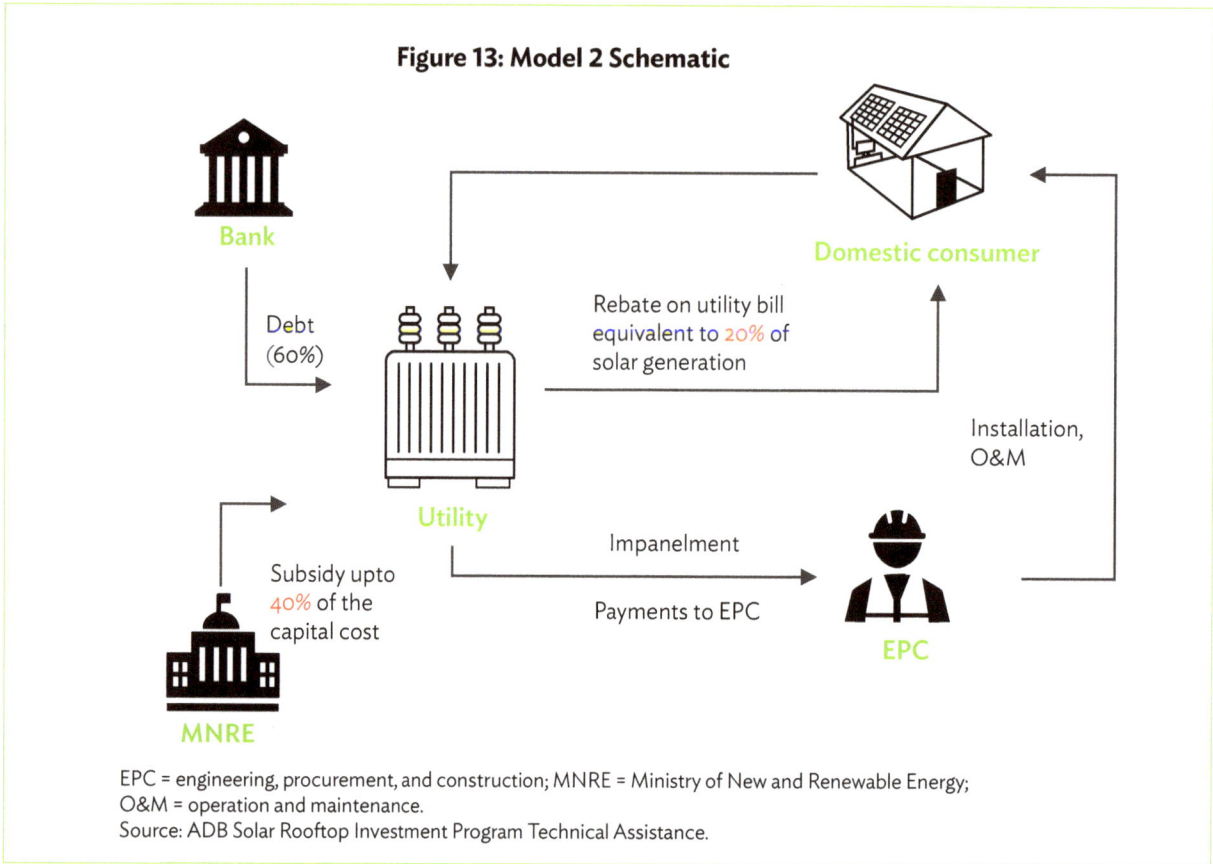

1. Target Consumer Category

Under this model, the rebate to the consumer is given in terms of units of electricity. Therefore, for the same number of units, revenue loss from high-paying consumers will be higher than the revenue loss from low-paying consumers. Therefore, Model 2 is recommended for the domestic consumer category.

2. Roles and Responsibilities of Stakeholders

Table 15 presents the roles and responsibilities of various stakeholders under Model 2.

Table 15: Roles and Responsibilities of Stakeholders

Consumer	Utility	EPC
(i) Make roof-space available.	(i) Aggregate interest from the consumers.	(i) Undertake designing, supply, installation, metering, testing, and commissioning.
(ii) Enter into an interconnection agreement with the utility.	(ii) Avail of and facilitate subsidy from the MNRE.	(ii) Perform installations and allied works till the interconnection point.
(iii) Agree to a lock-in period for 5 years from the date of commissioning.	(iii) Invest the remaining capital contribution.	(iii) Provide O&M for 5 years from the date of project commissioning.
(iv) Ensure safety and security of the plant.	(iv) Select the EPC through competitive bidding.	
	(v) Monitor the installation of the system on the consumer's rooftop.	

EPC = engineering, procurement, and construction; MNRE = Ministry of New and Renewable Energy; O&M = operation and maintenance.

Source: ADB Solar Rooftop Investment Program Technical Assistance.

3. Financial Impact on Stakeholders

Figure 14 presents cost benefit analysis for the consumer and the utility.

Figure 14: Cost–Benefit Analysis for Consumers and Utility

	Consumer	Utility
COSTS	• None	• Capital investment • Rebate to the consumer in terms of energy credits
BENEFITS	• No capital investment • Savings on utility bill based on the electricity generated from GRPV	• Reduction in T&D losses • Savings in power purchase cost from marginal power plants • RPO benefit, if applicable • Incentives and service charges under MNRE Phase II rooftop solar scheme

GRPV = grid-connected rooftop solar photovoltaic, O&M = operation and maintenance, MNRE = Ministry of New and Renewable Energy, RPO = renewable purchase obligation, T&D = transmission and distribution.
Source: ADB Solar Rooftop Investment Program Technical Assistance

4. Assumptions for Cost–Benefit Assessment

The following parameters are assessed to understand the benefit to utilities:

- loss of revenue to utility (ABR and revenue realization);
- benefit of savings in power purchase at APPC;
- savings in distribution losses due to localized consumption of power;
- avoidance of REC purchase by those utilities which are not meeting RPO targets; and
- rebate to the consumer considering the return on investment at 14%.

To assess the benefit to the utility, a case of Kerala State Electricity Board Limited (KSEBL) is presented, considering an individual project size of 3kWp. Table 16 presents the state-specific assumptions considered for calculating the benefit to the utility.

Table 16: Assumptions for Calculation of Benefit to Utility

Parameters	Value
Average revenue realized (low tension domestic)	₹4.62 per kWh
Escalation in utility tariff	2.00%
Variable charge rate of marginal plants	₹4.36 per kWh
Transmission and distribution losses	11.62%
Variable charge rate of marginal plants including transmission and distribution losses	₹4.87 per kWh
Escalation of variable charges of marginal power plants	2%
Renewable energy certificate floor price	₹1.00 per kWh
Rebate on utility bill equivalent to the percentage of solar generation	20%

kWh = kilowatt-hour.

Notes: 1. Parameters are assumed based on the Aggregate Revenue Requirement order of the Kerala State Electricity Board for FY2022. 2. The cost of generation for RTS plant is calculated based on the assumptions presented in Tables 4 and 6.

Sources: 1. Kerala State Electricity Board. 2019. Aggregate Revenue Requirement Order of the Kerala State Electricity Board for FY2022. https://www.kseb.in/index.php?option=com_jdownloads&task=download.send&id=10968&catid=64&m=0&lang=en. 2. ADB Solar Rooftop Investment Program Technical Assistance.

5. Benefit to the Utility

Figure 15 shows the benefit on a yearly basis for RTS plants aggregating to 1MW installed on the roof of domestic consumers.

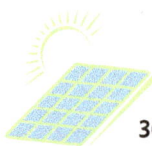

Figure 15: Benefit to Utility for a Typical 1 Megawatt Aggregated Capacity

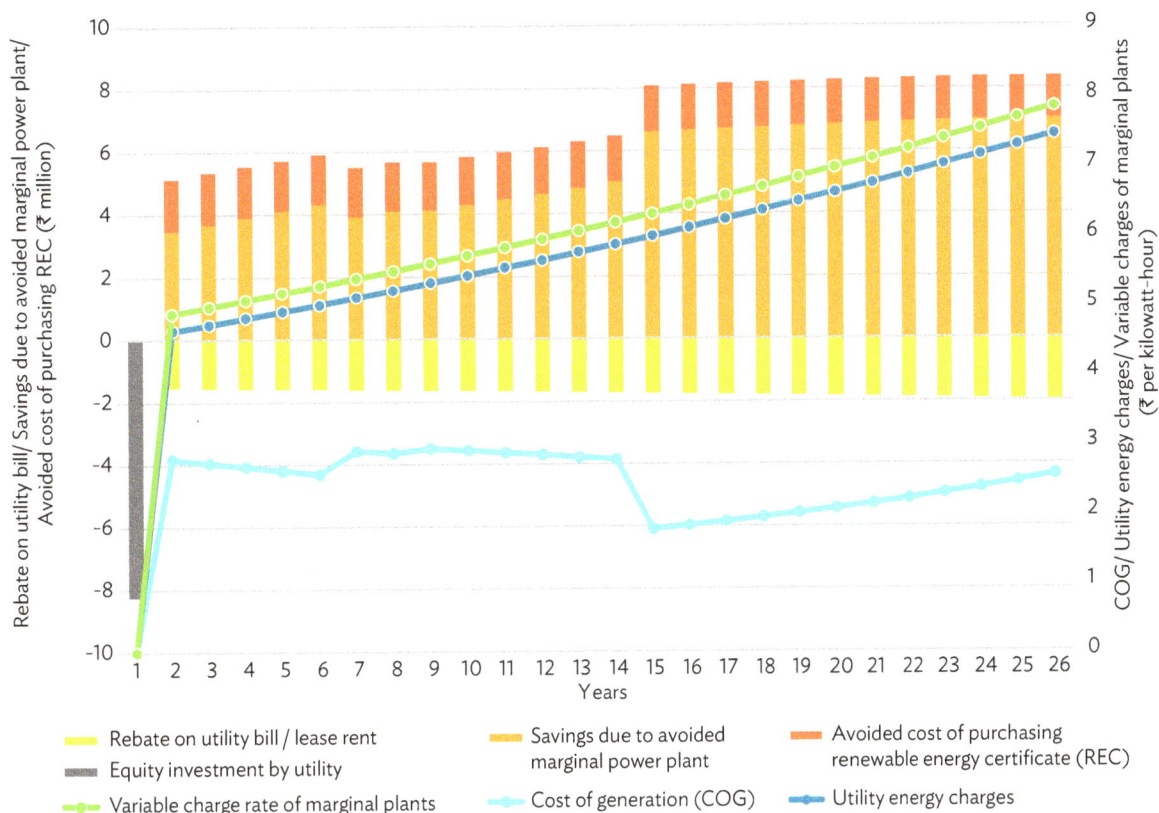

Source: ADB Solar Rooftop Investment Program Technical Assistance.

Table 17 shows the parameters of cost–benefit for the utility on per unit basis levelized at a discount rate of 8.5%.

Table 17: Cost–Benefit Analysis Parameters

Parameters	Levelized cost per unit
Marginal power purchase cost	5.74
Cost of generation	−2.59
Gross savings	3.15
Benefit of renewable purchase obligation (equivalent to renewable energy certificates)	1.00
Rebate to consumer	−1.09
Net savings	3.06

Source: ADB Solar Rooftop Investment Program Technical Assistance.

(i) In the above case, the utility earns up to ₹3.06 per kWh on each unit of power generation from the RTS.

(ii) For the consumer category with lower energy charges, benefit to the utility increases, as the rebate value is lower.

(iii) In cases where utilities have such high returns, high rebate percentages can be considered for increasing the benefit to the consumer as well.

(iv) As the utility is investing in this business model, approval from the State Electricity Regulatory Commission may be required.

(v) In addition, the utility will be eligible for incentives under the MNRE Phase II grid-connected RTS scheme based on the installed capacity in the previous year and capacity addition during the current year.

6. Benefit to the Consumer

Under this business model, there is no investment by the consumer. Rather, utilization of roof utility provides a rebate to the consumer on the utility bill. Table 18 shows the benefit to the consumer for the installation of 3kWp and rebate in the form of energy credits of 20% of the RTS generation.

Table 18: Benefit to the Consumer for 3 Kilowatt-peak Rooftop Solar Plant

Parameters	Unit	Value
Rebate to consumer	%	20
Average per month units saved on utility bill over 25 years	kilowatt hour	74
Average monthly rebate on utility bill over 25 years	`	434

Source: ADB Solar Rooftop Investment Program Technical Assistance.

7. Key Features of Model 2

(i) The utility invests and procures all the generated power from the RTS system, and the consumers receive a rebate in lieu of giving the utility the right to use their roof.

(ii) Utility avails of and facilitates subsidy from the MNRE.

(iii) The model is most appropriate for domestic and residential consumers.

Box 2: Soura Program, Kerala

The Government of Kerala launched the project Soura to add 1,000 megawatt-peak (MWp) solar power plants to the grid of Kerala State Electricity Board Limited (KSEBL), under the Urja Kerala Mission. Out of 1,000 MWp, 500 MWp was earmarked for rooftop solar (RTS) plants. The KSEBL is implementing the Soura program through the deployment of innovative utility-driven business models. The Soura project is India's first RTS program implemented with support and investment from a public sector distribution utility.

continued on next page

The KSEBL proposed three business models for the consumers to choose from for the implementation of the RTS projects.

(i) Roof-lease model

(ii) Master Renewable Energy Service Company (RESCO) model

(iii) Engineering Procurement and Construction (EPC) model

The consumers could choose any model based on the capital investment envisaged. The KSEBL received 278,000 consumer applications, out of which more than 75% opted for Model I. KSEBL tendered 200 MW to be implemented under Phase I of Soura, of which 150 MW was envisaged to be implemented by RESCO and 50 MW by EPC developers. The KSEBL engaged three EPC developers through a competitive bidding process for implementing 46.5 MW RTS projects under the above-mentioned business models. RESCOs did not submit bids for 150 MW considering smaller capacity of RTS projects.

Source: Study team analysis based on inputs received from Kerala State Electricity Board Limited.

C. Model 3: Engineering, Procurement, and Construction–Annuity Payment Basis Model

1. Model Outline

An EPC-annuity payment model involves partial contribution from all stakeholders as described below.

(i) **Consumer** contributes part of the capital required, based on which they receive rebate on the electricity bill in terms of energy credits equivalent to specific percentage of total generation from the RTS plant. The rebate increases with increased contribution from the consumer.

(ii) **Utility** invests the remaining capital and procures all the power generated from RTS plant at no cost.

(iii) **The RTS developer** installs the RTS and undertakes O&M for 5 years. The developer is paid two-thirds of the EPC cost at the time of commissioning of the plant and remaining one-third of the capital cost is paid on an annuity basis during the O&M contract.

This business model largely targets domestic consumers as RTS capacities in the domestic sector are small. Rooftop solar requires high capital cost and RESCOs do not prefer to implement such small individual projects (1–3 kilowatt) due to economic viability challenges. Therefore, it becomes difficult for low-paying consumers to install RTS projects. To address these challenges, Model 3 is proposed wherein initial capital cost is shared by the consumers and the utility, and thereafter the power generated by the RTS plant is similarly shared. The developer is paid part of the capital cost on an annuity basis for supervising the O&M during the contract period for 5 years (Figure 16).

Figure 16: Model 3 Schematic

Bank

Debt funding
(20% to 40%)

Capital contribution of
₹10,000 per kilowatt-peak

Domestic
consumer

Rebate on utility bill equivalent
to 35% of solar generation

Installation
and O&M

Subsidy up
to 40% of
the capital
cost

Utility

Selection through
competitive bidding

Payments to developer

Project
developer

MNRE

MNRE = Ministry of New and Renewable Energy, O&M = operation and maintenance.
Source: ADB Solar Rooftop Investment Program Technical Assistance.

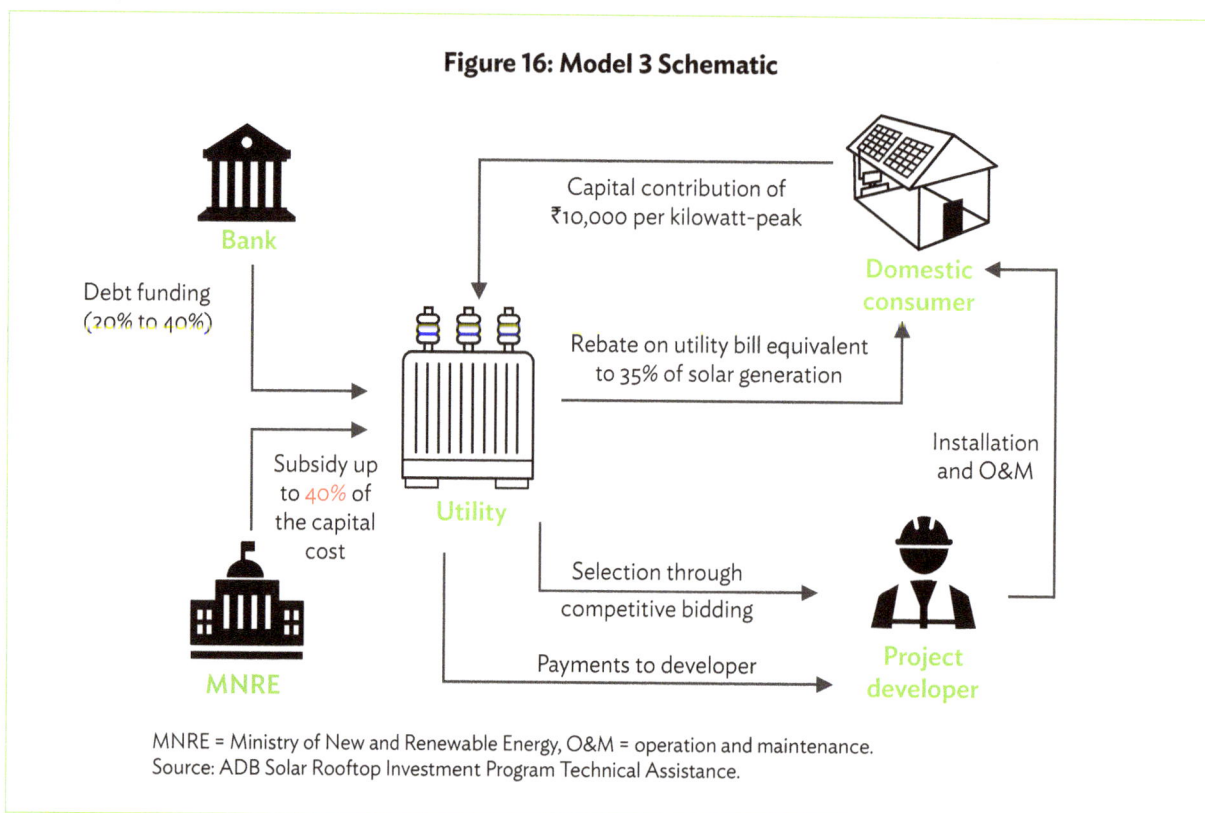

2. Investment Plan

The utility and the consumer both invest in the project. With increased investment from consumers, the rebate also increases. The developer is paid two-thirds of the capital cost at the time of commissioning of the RTS plant. Remaining one-third of the capital cost is paid to the developers on annuity basis over the O&M contract for 5 years. Table 19 describes the basis of assumption while determining tariff for the part payment to the developer.

Table 19: Stakeholder Contribution in Capital Cost for 1 Kilowatt-peak Rooftop Solar Plant

Contribution	Amount
Consumer contribution	₹10,000 per kWp
Utility contribution including subsidy	₹41,100 per kWp
Total capital expenditure (CAPEX)	₹51,100 per kWp

kWp = kilowatt-peak.
Source: ADB Solar Rooftop Investment Program Technical Assistance.

3. Roles and Responsibilities of Stakeholders

Table 20 presents the roles and responsibilities of various stakeholders under Model 3.

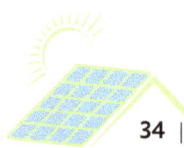

Table 20: Roles and Responsibilities of Stakeholders

Consumer	Utility	Developer
(i) Contribute part of the capital investment.	(i) Aggregate interest from the consumers.	(i) Undertake designing, supply, installation, metering, testing, and commissioning.
(ii) Make roof-space available.	(ii) Avail of and facilitate subsidy from MNRE, if applicable.	(ii) Perform installations and allied works up to the interconnection point.
(iii) Enter into an interconnection agreement with the utility.	(iii) Invest the remaining capital.	(iii) Provide O&M for 5 years from the date of project commissioning.
(iv) Ensure safety and security of the plant.	(iv) Select the project developer through competitive bidding.	
	(v) Monitor the installation of the system on the consumer's rooftop.	

MNRE = Ministry of New and Renewable Energy; O&M = operation and maintenance.
Source: ADB Solar Rooftop Investment Program Technical Assistance.

4. Financial Impact on Stakeholders

Figure 17 represents assessment of cost–benefit analysis for the consumers and the utility.

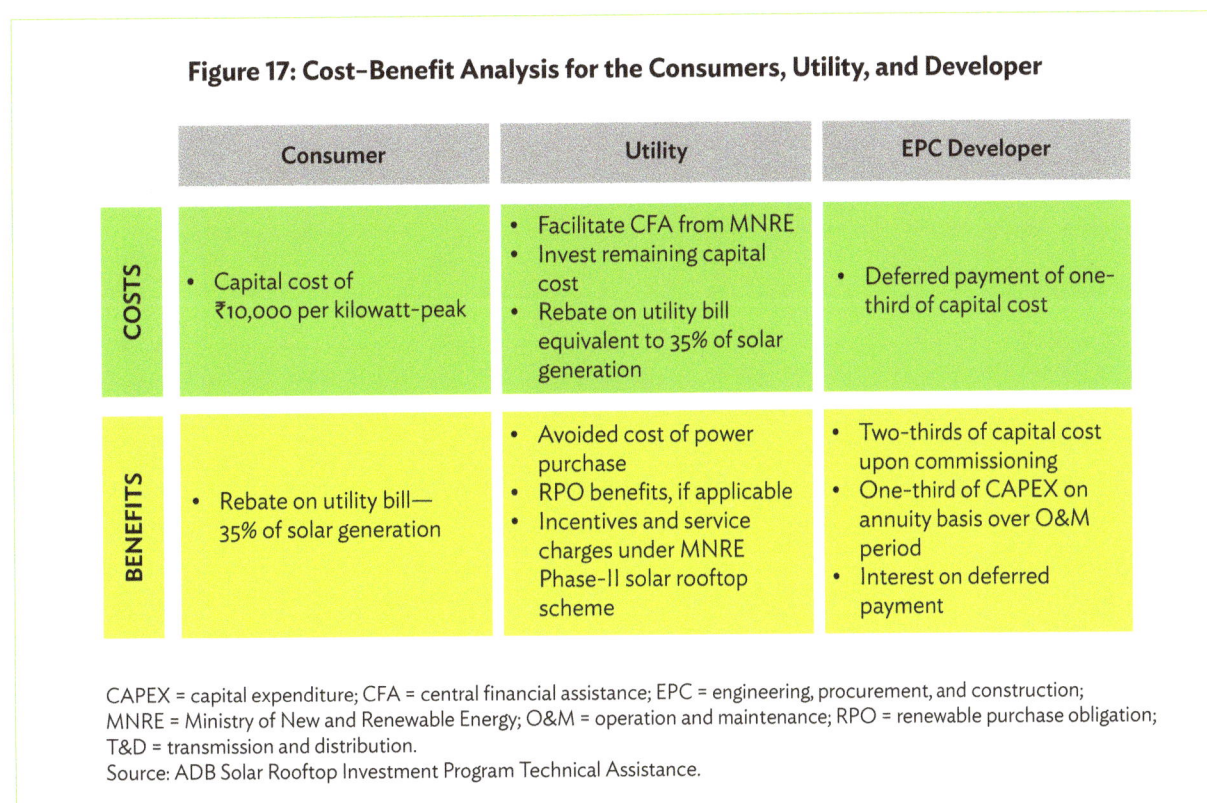

Figure 17: Cost–Benefit Analysis for the Consumers, Utility, and Developer

	Consumer	Utility	EPC Developer
COSTS	• Capital cost of ₹10,000 per kilowatt-peak	• Facilitate CFA from MNRE • Invest remaining capital cost • Rebate on utility bill equivalent to 35% of solar generation	• Deferred payment of one-third of capital cost
BENEFITS	• Rebate on utility bill—35% of solar generation	• Avoided cost of power purchase • RPO benefits, if applicable • Incentives and service charges under MNRE Phase-II solar rooftop scheme	• Two-thirds of capital cost upon commissioning • One-third of CAPEX on annuity basis over O&M period • Interest on deferred payment

CAPEX = capital expenditure; CFA = central financial assistance; EPC = engineering, procurement, and construction; MNRE = Ministry of New and Renewable Energy; O&M = operation and maintenance; RPO = renewable purchase obligation; T&D = transmission and distribution.
Source: ADB Solar Rooftop Investment Program Technical Assistance.

5. Assumptions for the Cost–Benefit Assessment

The following parameters are to be assessed to understand the benefit to utilities:

    (i) loss of revenue to utility (ABR or revenue realization);

    (ii) benefit of savings in power purchase considered at APPC;

(iii) savings in distribution losses due to localized consumption of power; and

(iv) avoidance of REC purchase by those utilities which are not meeting RPO targets.

To assess the benefit to the utility, a case of KSEBL is presented. Table 21 illustrates the assumptions for power generation from the RTS plant.

Table 21: State-Specific Assumptions for Assessing Benefits to the Utility

Parameters	Value
Average revenue realized	₹4.62 per kWh
Escalation in utility tariff	2.00%
Variable charge rate of marginal plants	₹4.36 per kWh
Transmission and distribution losses	11.62%
Variable charge rate of marginal plants including transmission and distribution losses	₹4.87 per kWh
Escalation of variable charges of marginal power plants	2.00%
Renewable energy certificate floor price	₹1.00 per kWh
Rebate on utility bill equivalent to the percentage of solar generation	35.00%

kWh = kilowatt-hour.

Note: Parameters are assumed based on the Aggregate Revenue Requirement order of the Kerala State Electricity Board for FY2022.

Sources: 1. Kerala State Electricity Board. 2019. Aggregate Revenue Requirement Order of the Kerala State Electricity Board for FY2022. https://www.kseb.in/index.php?option=com_jdownloads&task=download.send&id=10968&catid=64&m=0&lang=en.
2. ADB Solar Rooftop Investment Program Technical Assistance.

The cost of generation of the RTS plant is calculated based on the assumptions considered in Table 6.

In addition, the following business-model specific assumptions are considered:

(i) Consumers invest ₹10,000 per kWp and the remaining investment is made by the utility inclusive of subsidy. The transaction between the utility and the consumer under Model 3 is similar to Model 2. In Model 2, no investment was made by the consumer and a fixed rebate was available. With consumer's investment, a higher rebate is offered.

(ii) The utility pays the developer two-thirds of the total capital cost at the commissioning of the RTS plant. Remaining one-third of the capital cost is paid on an annuity basis over the period of O&M contract which is generally for 5 years.

(iii) The cost of generation for the utility is calculated for the utility contribution in capital cost (₹41,100 per kWp inclusive of subsidy).

6. Benefit to the Utility

Based on the above assumptions, the benefit to utility is calculated as:

Net Benefit to DISCOMs
= (Average power purchase cost
– Cost of generation from rooftop solar for utility capital contribution)
– *Rebate to consumer* – Deferred payment for one-third of capital cost including interest + *Avoided cost of purchasing RECs*

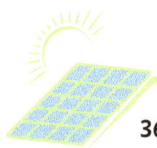

Figure 18 shows the benefit to utility for the installation of an RTS of aggregated capacity of 1MW on a yearly basis for the domestic consumer category. An individual plant capacity for 3 kWp is considered at a capital cost.

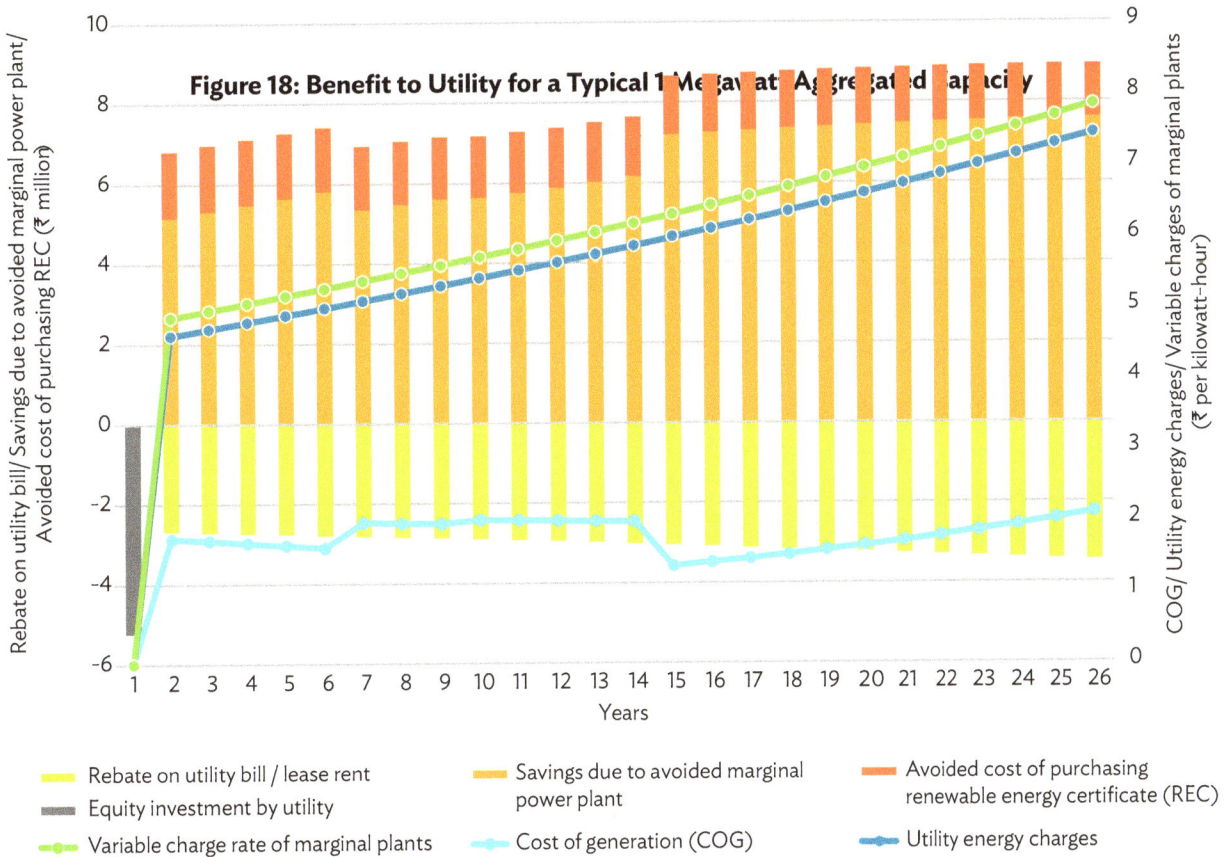

Figure 18: Benefit to Utility for a Typical 1 Megawatt Aggregated Capacity

Legend:
- Rebate on utility bill / lease rent
- Savings due to avoided marginal power plant
- Avoided cost of purchasing renewable energy certificate (REC)
- Equity investment by utility
- Variable charge rate of marginal plants
- Cost of generation (COG)
- Utility energy charges

Source: ADB Solar Rooftop Investment Program Technical Assistance.

Table 22 shows the parameters of cost–benefit for the utility on a per unit basis levelized at a discount rate of 8.5%.

Table 22: Parameters of Cost–Benefit for the Utility

Parameters	Levelized cost per unit (₹)
Avoided marginal power plant cost	5.74
Cost of generation	−1.79
Gross savings	3.95
Benefit of renewable power obligation (equivalent to renewable energy certificate)	1.00
Rebate to consumer	−1.91
One-third of engineering, procurement, and construction cost paid to the developer	−1.87
Net savings	1.17

Source: ADB Solar Rooftop Investment Program Technical Assistance.

(i) In Model 3, the utility gains ₹1.17 per kWh of power generated from the RTS plant.

(ii) Especially for small size domestic RTS plants, adopting this model will ensure regular O&M of the plant.

(iii) Part-contribution by all stakeholders reduces risk for each stakeholder thus improving the financial viability of the project.

(iv) Buy-in from all the stakeholders is necessary for the successful implementation of this business model.

(v) The utility is also eligible for incentives under the MNRE Phase II of grid-connected RTS scheme.

7. Benefit to the Consumers

Under Model 3, when the consumer invests ₹10,000 per kWp, the utility provides a rebate on the power bill equivalent to 35% of the energy generated from RTS plant. Table 23 shows the benefit to consumer under this business model.

Table 23: Return on Investment for 3 Kilowatt-peak Rooftop Solar Plant for Consumers

Parameter	Unit	Value
Capital cost	₹	1,37,280
Investment by consumer	₹	30,000
Rebate to consumer	%	35
Average per month units saved on utility bill over 25 years	kWh	130
Average monthly rebate on utility bill over 25 years	₹	760
Payback period	Years	3.3

kWh = kilowatt-hour.
Source: ADB Solar Rooftop Investment Program Technical Assistance.

The benefit both in terms of power and money value is averaged for 25 years. Payback for consumer is around 3.3 years for the part contribution in the capital cost.

8. Key Features of Model 3

(i) The utility and the consumer both invest in the project.

(ii) The utility engages the EPC for project installation and operations.

(iii) The utility avails of and facilitates the subsidy disbursement.

(iv) This model is highly relevant to small domestic and residential consumers with a monthly consumption of up to 200 kWh.

Box 3: Example of Developer Annuity Payment Model

The Ministry of New and Renewable Energy (MNRE), Government of India allocated 50 megawatts (MW) of capacity to Kerala State Electricity Board Limited (KSEBL) during the financial year (FY) 2020 and 200 MW in FY2021 for the implementation of rooftop solar (RTS) projects among residential consumers under the central financial assistance (CFA) scheme. To roll out the scheme according to the Phase II guidelines of the MNRE, the KSEBL devised three innovative business models considering the following drivers:

• Subsidy should be targeted at low-end domestic consumers with average monthly electricity consumption up to 200 kilowatt-hour (kWh).

continued on next page

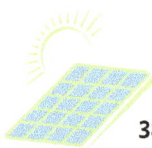

- All stakeholders should contribute to and have stake in the development of RTS projects.
- Interests of the utility, the consumers, and the developer should be protected through proper sharing of risks and rewards.
- The model scheme should be compatible with the regulatory framework prevalent in the state.

Model 1A for domestic consumers with a monthly consumption of up to 120 kWh. Customer cost share will be 12% of bid discovered price per kilowatt-peak (kWp) or ₹6,200 per kWp whichever is lower.

Share of Stakeholders (₹)						
Plant capacity	Total project cost	MNRE CFA	Consumer contribution	KSEBL investment	Developer investment	Energy share – consumer
2 kWp	1,08,000	43,200	12,400	16,400	36,000	25%
3 kWp	1,62,000	64,800	18,600	24,600	54,000	25%

Model 1B for domestic consumers with a monthly consumption of up to 150 kWh. Customer cost share will be 20% of bid discovered price per kWp or ₹11,000 per kWp, whichever is lower.

Share of Stakeholders (₹)						
Plant capacity	Total project cost	MNRE CFA	Consumer contribution	KSEBL investment	Developer investment	Energy share – consumer
2 kWp	1,08,000	43,200	22,000	6,800	36,000	40%
3 kWp	1,62,000	64,800	33,000	10,200	54,000	40%

Model 1C for domestic consumers with a monthly consumption up to 200 kWh. Customer cost share will be 25% of bid discovered price per kWp or ₹14,000 per kWp, whichever is lower.

Share of Stakeholders (₹)						
Plant capacity	Total project cost	MNRE CFA	Consumer contribution	KSEBL investment	Developer investment	Energy share – consumer
2 kWp	1,08,000	43,200	28,000	800	36,000	50%
3 kWp	1,62,000	64,800	42,000	1,200	54,000	50%

The payment to the solar power developer shall be as follows:

(i) Of the project cost discovered through the e-tender on EPC basis, 66.66% will be paid upon the completion of the project.

(ii) Of the project cost, 33.33% of the will be paid on annuity basis over a period of 7 years.

continued on next page

The KSEBL successfully impaneled three vendors for the implementation of 50 MW of the RTS project under the CFA scheme. Consumers have option of choosing between the conventional EPC model or the EPC annuity payment model. Of the 50 MW envisaged, around 2,600 applications aggregating to 6.8 MW were received as of December 2020, for the implementation of the RTS project under the EPC annuity model. Following prices were discovered through the vendor impanelment process.

Plant capacity	Price discovered (₹ per kWp)
2 kWp	42,999
3 kWp	41,998

Source: Study team analysis based on inputs received from Kerala State Electricity Board Limited.

D. Model 4: Utility as Master Renewable Energy Service Company

1. Model Outline

Under Model 4, the utility acts as a RESCO, installs the RTS and collects tariff from the consumer for the consumption of power from the RTS plant at a predetermined rate.

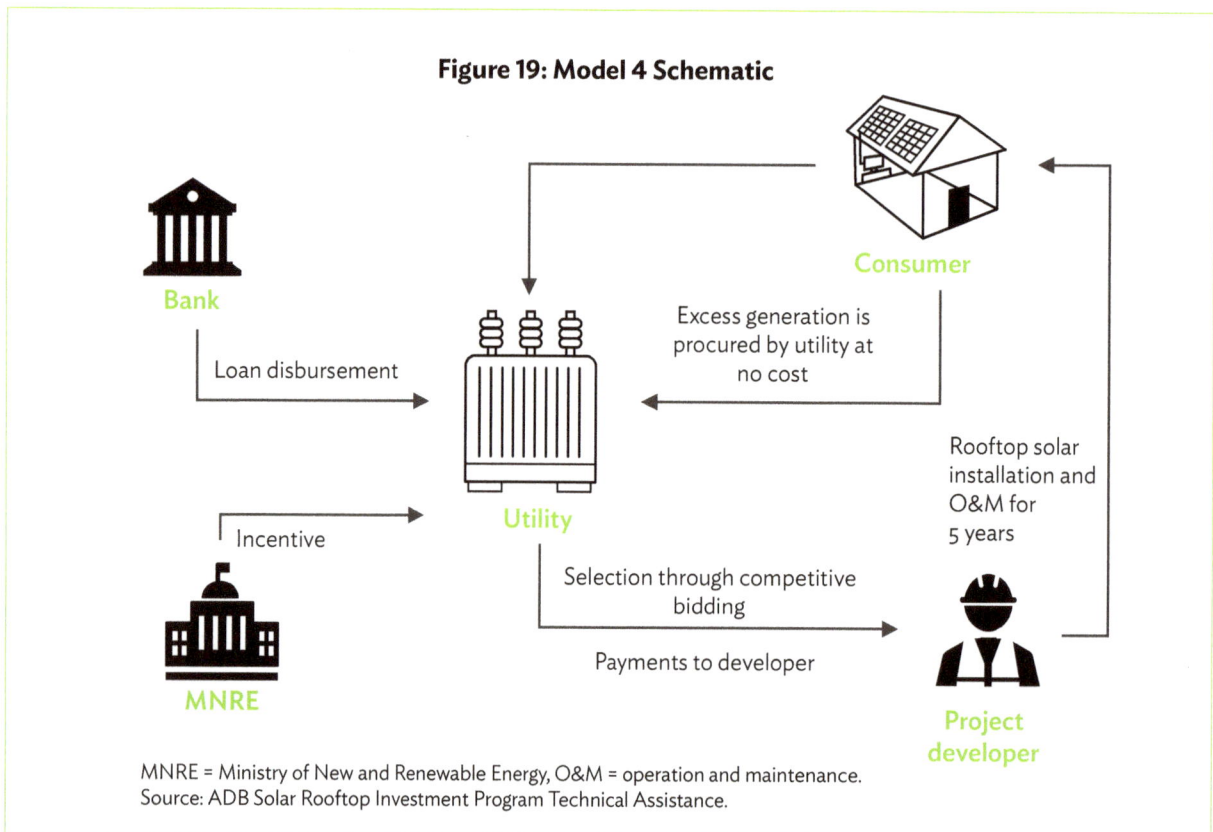

Figure 19: Model 4 Schematic

MNRE = Ministry of New and Renewable Energy, O&M = operation and maintenance.
Source: ADB Solar Rooftop Investment Program Technical Assistance.

The purpose of Model 4 is to retain high-paying (industrial and commercial) consumers, for whom solar photovoltaic works out cheaper than grid supply. At the same time, high-paying consumers constitute the main source of revenue for utilities, cross-subsidizing low-paying consumers. It is important, therefore, for the utilities to retain the high-paying consumer base to continue to serve low-paying consumers. Model 4 is proposed for the utility to invest in and install for the RTS project and sell its power to high-paying consumers at a tariff which is 10%–20% lower than the grid tariff. The utility can reduce the tariff based on the savings in power purchase cost with respect to the RTS cost of generation and distribution losses.

The consumer benefits from the reduced tariff and continued reliable power supply while the utility benefits by retaining the high-paying consumer, savings in power purchase from the marginal power plant, and shifting to a cheaper power source.

The business model can also be implemented through RESCO developers selected by the utility through competitive bidding. Power is purchased from the RESCO developer and sold to the consumer at a predetermined rate. The utility acts as a trader between the developer and the consumer. The purpose of keeping utility as a trader ensures quality of O&M over the life of the project as well as payment security to the developer. The utility thereby continues to retain its high-paying consumer base.

The cost of sale of power to the consumer is determined in accordance with the Central Electricity Regulatory Commission, which stipulates trading license adhering to the trading margins of not more than ₹0.07 per kWh for sale price exceeding ₹3.00 per kWh and not more than ₹0.04 per kWh for sale price less than or equal to ₹3.00 per kWh. The regulation restricts the trading margin for all trading licenses. In such a case, the tariff on solar power consumed by high-paying consumers is equal to the discovered RESCO tariff plus ₹0.07 per kWh, not exceeding ₹5.00 per kWp, considering the recent tariffs discovered through the bidding process for RESCO projects.

Under RESCO mode of implementation, though the utility can retain its consumer base, it will not be able to make enough revenue to balance the cross subsidization to other consumer category at this tariff. Therefore, it is suggested that the model be executed through EPC, where the utility invests in and installs the RTS on the roof of the high-paying consumer and sells the power at 10% to 30% lower than the retail tariff applicable for those consumer categories, based on the viability of the RTS project.

2. Role and Responsibilities of Stakeholders

Table 24 presents the roles and responsibilities of various stakeholders under Model 4.

Table 24: Roles and Responsibilities of Stakeholders

Consumer	Utility	EPC
(i) Enter into a power purchase agreement with utility. (ii) Ensure safety and security of the system.	(i) Invest capital cost for the project. (ii) Select project EPC through competitive bidding. (iii) Monitor the installation of the system on the consumer's roof. (iv) Operate and maintain the system.	(i) Undertake designing, supply, installation, metering, testing, and commissioning. (ii) Perform installations and allied works up to the interconnection point. (iii) Provide O&M for 5 years from the date of project commissioning.

EPC = engineering, procurement, and construction; O&M = operation and maintenance.
Source: ADB Solar Rooftop Investment Program Technical Assistance.

3. Financial Impact on Stakeholders

Figure 20 presents the cost–benefit analysis for consumers and the utility.

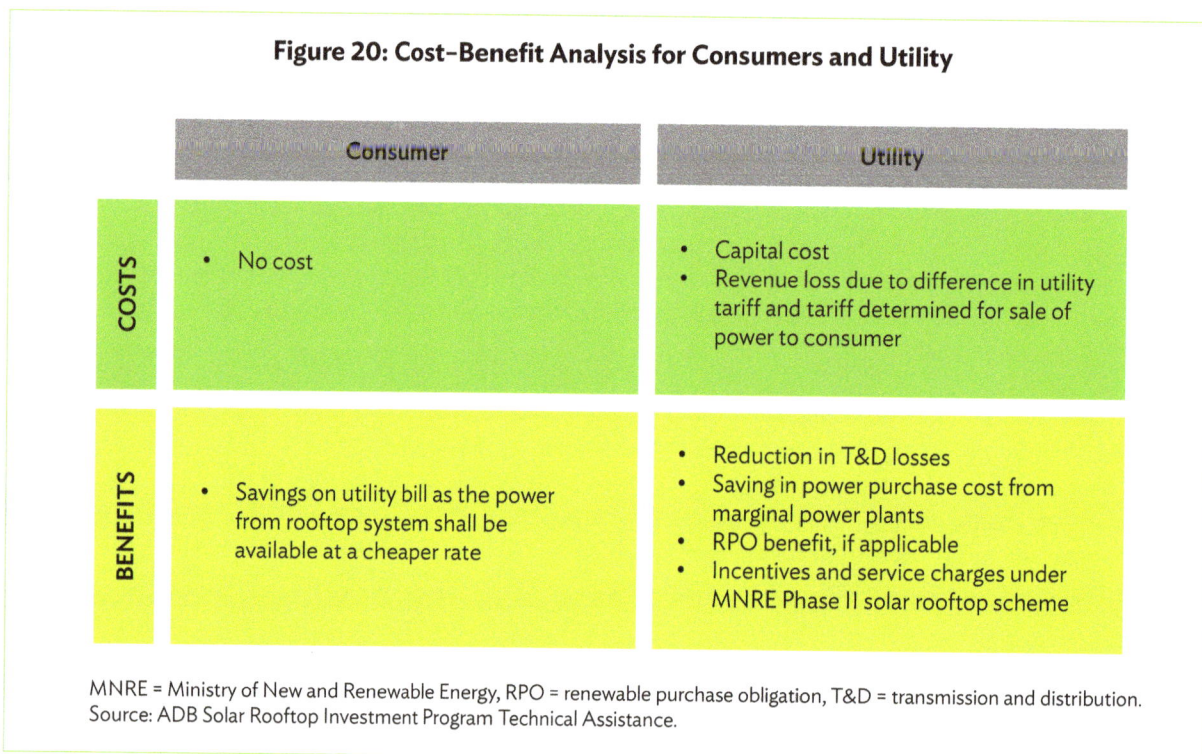

Figure 20: Cost–Benefit Analysis for Consumers and Utility

	Consumer	Utility
COSTS	• No cost	• Capital cost • Revenue loss due to difference in utility tariff and tariff determined for sale of power to consumer
BENEFITS	• Savings on utility bill as the power from rooftop system shall be available at a cheaper rate	• Reduction in T&D losses • Saving in power purchase cost from marginal power plants • RPO benefit, if applicable • Incentives and service charges under MNRE Phase II solar rooftop scheme

MNRE = Ministry of New and Renewable Energy, RPO = renewable purchase obligation, T&D = transmission and distribution.
Source: ADB Solar Rooftop Investment Program Technical Assistance.

4. Assumptions for Cost–Benefit Assessment

The following parameters are assessed to understand the benefit to utilities:
- (i) loss of revenue to utility (ABR and revenue realization);
- (ii) benefit of savings in power purchase considered at APPC;
- (iii) savings in distribution losses due to localized consumption of power;
- (iv) avoidance of REC purchase by those utilities not meeting RPO targets; and
- (v) cost of generation from the RTS plant.

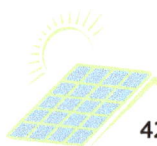

Table 25: State-Specific Assumptions for Estimation of Benefit to Utility

Parameters	Value
Average revenue realized from low-tension industrial consumer	₹10.11 per kWh
Escalation in utility tariff	2%
APPC inclusive of T&D losses	₹6.30 per kWh
Escalation in APPC	2.00%
REC floor price	₹1.00 per kWh
Tariff determined for sale of solar power to consumer (15% lower than utility retail tariff)	₹9.09 per kWh

APPC = average power purchase cost, kWh = kilowatt-hour, REC = renewable energy certificate, T&D = transmission and distribution.

Notes: 1. Parameters are assumed based on the Aggregate Revenue Requirement order of the Karnataka Electricity Regulatory Commission for Bangalore Electricity Supply Company for FY2022.
2. In Karnataka, power procurement allocation is on pooled basis where aggregated power procured is distributed across all the distribution companies of Karnataka. Hence, APPC is considered.
3. The cost of generation for the rooftop solar plant is calculated based on the assumptions presented in Tables 4 and 6.

Sources: 1. Karnataka Electricity Regulatory Commission. Aggregate Revenue Requirement order of the Karnataka Electricity Regulatory Commission for Bangalore Electricity Supply Company for FY2022. https://kerc.karnataka.gov.in/storage/repo/Chapter-4_Modified%20ARR%20 for%20FY-22.docx.
2. ADB Solar Rooftop Investment Program Technical Assistance.

To assess the benefit to the utility, a case of Bangalore Electricity Supply Company Limited (BESCOM) is presented. Table 25 depicts the estimation of benefit to utility.

In addition, the following business-model specific assumptions are considered:

(i) Under Model 4, the impact on the utility of the installation of the RTS on rooftops of high-paying consumers is evaluated. Since industrial and commercial consumers are considered, RTS capacity of 10kWp–100kWp is assumed. The capital cost for the selected RTS capacity group is ₹41,640 per kWp.

(ii) Impact is calculated of the installation of RTS on rooftops of high-tension industrial consumers in the BESCOM area.

(iii) Tariff for sale of power generated from the RTS plant is considered to be 15% lower than the utility tariff, with the percentage varying with the expected return on investment for the utility.

(iv) BESCOM has achieved its RPO targets and, therefore, benefits from savings on REC purchase are not considered in the calculation.

5. Benefit to the Utility

Based on the above assumptions, benefit to the utility is calculated using the following formula:

Net Benefit to BESCOMs
= (Average power purchase cost
– Cost of generation from rooftop solar for utility capital contribution)
– *Loss in utility revenue* – Revenue from sale of solar power to consumer
+ *Avoided cost of purchasing RECs*

Figure 21 shows the benefit on a yearly basis for the RTS plants aggregating to 1MW installed on the roofs of high-tension industrial consumers in the BESCOM area.

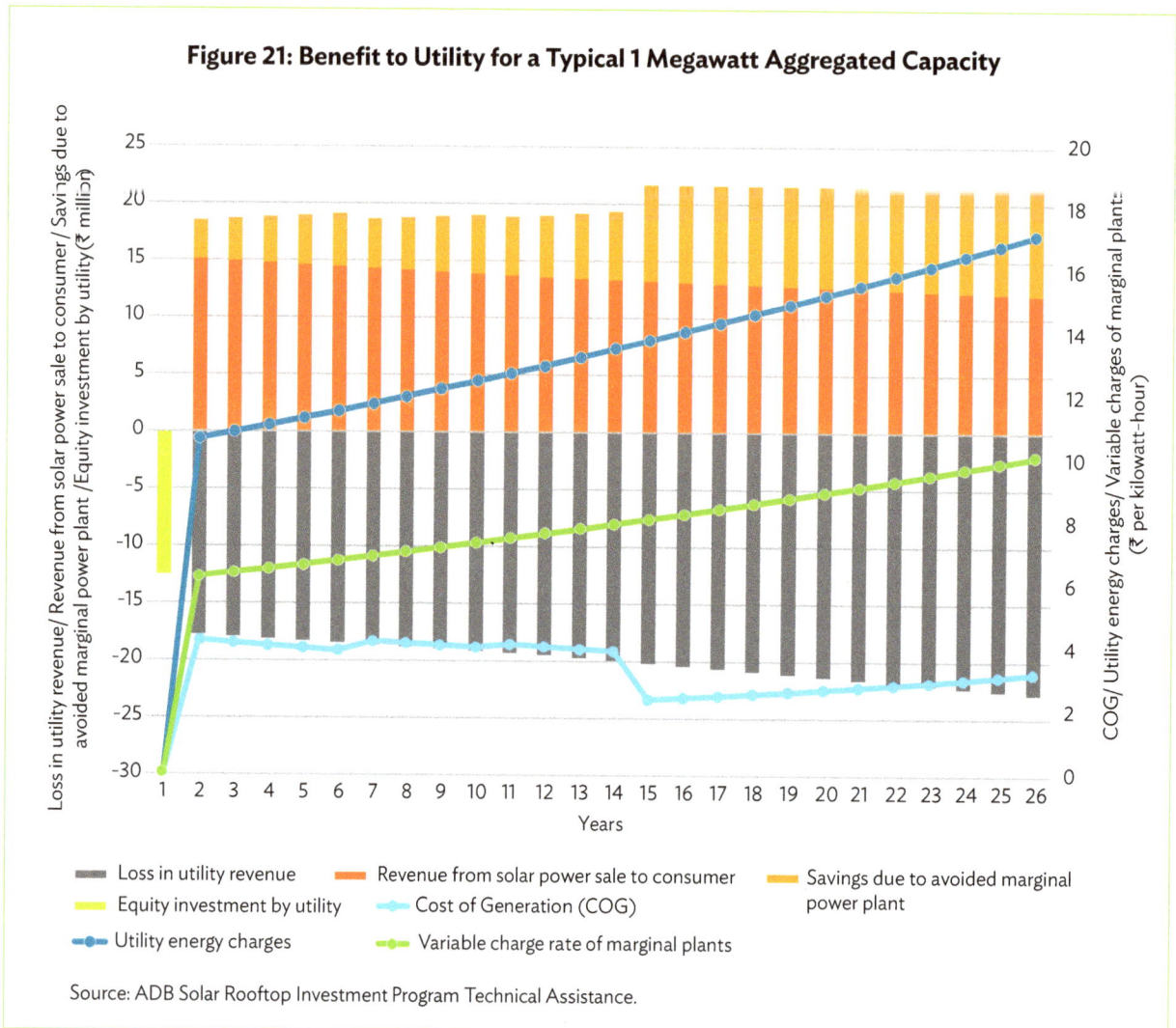

Figure 21: Benefit to Utility for a Typical 1 Megawatt Aggregated Capacity

Source: ADB Solar Rooftop Investment Program Technical Assistance.

Table 26 shows the parameters of cost–benefit for the utility per unit levelized cost at a discount rate of 8.5%. Savings on the purchase of RECs is not considered as BESCOM already complies with RPO targets.

Table 26: Parameters of Cost–Benefit for the Utility

Parameters	Levelized cost per unit (₹)
Avoided marginal power plant cost	7.44
Cost of generation	–3.78
Gross savings	3.66
Benefit of RPO (equivalent REC)	0
Loss of utility revenue	–12.62
Revenue from sale of solar power to consumer	9.09
Net savings	0.13

REC = renewable energy certificate, RPO = renewable purchase obligation.
Source: ADB Solar Rooftop Investment Program Technical Assistance.

(i) Even after loss of revenue due to sale of power at reduced tariff as compared to retail supply tariff, utility benefit is around ₹0.13 per kWh on each unit generated from the RTS plant.

(ii) Utility shall retain its high paying consumer base which is a major source of revenue for utilities.

6. Benefit to Consumer

Consumers do not invest in this business model; they sign a power purchase agreement (PPA) with the utility, and receive power generated by the RTS plant at a predetermined tariff. Such consumers enjoy these benefits:

(i) Power is available at a lower tariff. Tariff through the PPA tenure remains fixed, whereas grid tariff is expected to increase annually.

(ii) Grid supply is reliable.

(iii) As a developer, the utility guarantees service for the project's lifetime.

(iv) The standardized RTS system ensures quality and regular O&M by the utility.

7. Key Features of Model 4

i) Under the master RESCO model, the utility acts as a RESCO and installs the RTS on the consumer's roof and collects tariff from the consumer for the consumption of power from the RTS plant at a predetermined tariff.

ii) The model is beneficial for high-end consumers, as they receive supply at a lower tariff, fixed for 25 years.

iii) The utility gains through this model by retaining high-paying consumers. Such high loads increase over time, leading to an increase in electricity demand from the utility.

6 SUMMARY OF BUSINESS MODELS

Table 27: Summary of Business Models

	Model 1	Model 2	Model 3	Model 4
Key driver	Consumer-owned with utility as facilitator	Roof-leasing with utility investment	EPC on an annuity basis with partial stakeholder contribution	Utility as master Renewable Energy Service Company
Target consumer category	All consumer categories	Domestic consumers	Domestic consumers up to 200 kWh monthly consumption	High paying consumers, namely, industrial and commercial
Model of development	EPC	EPC	EPC on an annuity basis	EPC
Key consideration	Utility as a facilitator and aggregator can target specific areas based on high distribution losses or poor billing efficiency.	The business model is beneficial for the utility for all consumer categories, but benefit is higher for consumers with low energy charges.	Business model targets appropriate risk and reward sharing mechanism across all stakeholders.	Higher the gap between average power purchase cost and consumer tariff, higher will be the benefits for the utility.
Consumer: (i) Investment (ii) Benefit (iii) Payback	(i) Complete contribution, inclusive of debt (ii) Reduction in electricity bill (iii) 5–7 years	(i) No investment (ii) 10% energy credit of gross generation (iii) Not applicable	(i) Partial investment (ii) ₹10,000 per kW (approximately 18% of cost); 35% energy credit of gross generation (iii) 4–7 years	(i) No investment (ii) Power available at tariff lower than grid tariff (iii) Not applicable
Utility: (i) Investment (ii) Benefit (iii) Payback	(i) No investment (ii) Savings in marginal power purchase	(i) Complete investment (ii) Savings in marginal power purchase and consumer retention (iii) 4–7 years	(i) Remaining capital investment (ii) 20%–42% of investment; savings in marginal power purchase and consumer retention (iii) 4–7 years	(i) Complete capital investment (ii) Savings in marginal power purchase and high paying consumer retention (iii) 10–12 years
Role of developer	EPC and O&M for 5 years	EPC and O&M for 5 years	EPC along with one-third of capital cost recovered on annuity basis over 5 years	EPC with O&M for 5 years

EPC = engineering, procurement, and construction; kW = kilowatt; kWh = kilowatt-hour; O&M = operation and maintenance.
Source: ADB Solar Rooftop Investment Program Technical Assistance.

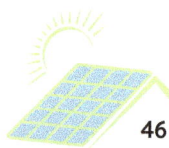

7 CONCLUSION

New and innovative business models can drive the RTS market. One of the major challenges is that distributing companies perceive RTS schemes to be causing loss of revenue. As distribution utilities are the primary agencies involved in approving RTS projects, it is essential to consider their interests. As a technology option, RTS reaching grid parity has led to increased adoption; but it is still on a very small scale when compared to the national demand. It is vital that utilities adapt to changing consumption and demand patterns, for the survival of their business.

Going for RTS offers many advantages to utilities—including reduced technical and commercial losses, better demand management, and savings in power procurement. Realizing these benefits, several utilities across various states have announced programs where they play active roles as aggregators.

Aiming to help utilities plan successful RTS projects in their service areas, this guidebook proposes four potential business models, each of which is designed to maximize economic benefits for utilities through the deployment of RTS, while avoiding the need for consumers to fund costly capital investments.

In this guidebook, we have provided examples of each business model, using broad assumptions. Adopting a similar cost-benefit analysis, utilities can evaluate the economic advantages of RTS deployment in their prevailing local circumstances. Such models can be adapted to target low-paying consumers—where the utility is highly dependent on cross-subsidy—or in areas where there is high distribution-transformer loading, or substantial distribution losses. Such targeted approaches will not only forestall the utility's need for immediate expenditure, but it will also ensure long-term benefits for all stakeholders.

APPENDIX: APPLICABLE METERING FRAMEWORK BY STATE

State	Applicable Metering Framework
Andhra Pradesh	• The projects of capacity up to 1,000 KWp at a single location will be permitted. • The consumer(s) are free to choose either net or gross meter option for sale of power to distribution company (DISCOM) under this policy.
Assam	• All eligible individual consumers can avail the facility of net metering and export–import (EXIM) metering for capacity of 1–1,000 kWp.
Bihar	• Eligible consumer can install the power plant under net or gross metering mechanism. • The capacity to be installed at the premises of any eligible consumer under these regulations shall not be less than 1 kilowatt peak (kWp) and shall not exceed the sanctioned or contracted load of the eligible consumer.
Chhattisgarh	• The capacity of the system shall not exceed the sanctioned load or contract demand of the prosumer. • Maximum size of renewable energy system that can be set up under net metering arrangement would be 500 kW.
Joint Electricity Regulatory Commission (JERC) of Goa	• Net and gross metering mechanisms are available for all consumers from 1 kW onward with no limit on maximum capacity. • Virtual and group net-metering is available from 5 kW onward with no limit on maximum capacity. • Virtual net metering framework is applicable for consumers under domestic category, consumers like hospitals, colleges, schools, other institutions run or managed by charitable institutions, non-profit organizations/trusts, which are not covered under the category of domestic consumers, offices of government/local authorities and renewable energy generators registered under Mukhya Mantri Kisaan Aay Badhotari Yojna.
Gujarat	• Net metering provision for projects with capacity of 1 kilowatt (kW) and above and up to 1,000 kW. • Gross metering provision for projects 10 kW–1,000 kW. • Residential consumers allowed irrespective of consumer sanctioned load. • No capacity restrictions up to sanctioned load/contracted demand is applicable for captive consumers and the project is set up under third party sale consumers within the permissible limit.
Haryana	• The eligible consumer may install the system under net or gross metering arrangement. • Net metering: 1 kW to max connected load (up to 500 kW). • Gross metering: 1 kW to max connected load (no limit)
Himachal Pradesh	• All domestic consumers of Himachal Pradesh State Electricity Board Limited shall be eligible to install grid-connected RTS irrespective of consumer sanctioned load.
Jharkhand	• The capacity to be installed by any eligible consumer or third-party owner shall be from 1 kWp to 2 MWp under net metering and gross metering mechanism.

State	Applicable Metering Framework
Karnataka	• The minimum system capacity should be 1 kW and maximum capacity could be up to 2,000 kW or sanctioned load whichever is lower. • The eligible consumers can install the systems under both net metering and gross metering mechanisms.
Kerala	• All the eligible consumers can install RTS systems of between 1 kW and 1 MW under net-metering mechanism.
Madhya Pradesh	• Eligible consumer may install RTS under net metering mechanism up to 500kWp, while under gross metering mechanism from 500kWp up to 1MW.
Maharashtra	• Net metering: 1–1,000 kW for all sectors. • Net billing: 1–1,000 kW for all sectors. • Behind-the-meter: 1–1,000 kW for all sectors.
JERC for Manipur and Mizoram	• All the eligible individual consumers are allowed to install RTS in the 1 kWp–1 MWp range capacity under net metering and gross metering mechanisms.
Meghalaya	• Net metering is available for all the consumers with 1 kW–1 MW installed capacity.
Nagaland	• All eligible consumers can avail the benefits of RTS systems for the capacity of 1– 500 kWp under net metering and gross metering mechanisms.
Odisha	• The capacity to be installed by any eligible consumer or third party owner under net metering shall be from 1 kWp to 500 kWp. • The capacity of the renewable energy system under group net metering or virtual net metering framework to be installed by any renewable energy generator shall not be less than 5 kW and more than 500 kW.
Punjab	• The maximum capacity of systems under net metering shall not exceed 500 kW. • The maximum capacity to be installed at any eligible consumer's premises except domestic category consumers, shall not exceed 70 % of the sanctioned load (kW) or contract demand of the consumer (in kVA converted to kW by using a power factor of 0.9). • The minimum capacity of system under net metering or net billing arrangements shall be 1 kWp for a single eligible consumer. • Under gross metering arrangements, the minimum capacity shall be 50 kWp for a single eligible consumer.
Rajasthan	• All eligible consumers of the distribution licensee having or proposing to install a renewable energy generating system may opt for grid connectivity under the net billing arrangement or net metering arrangement for system capacity from 1 kW up to the sanctioned load.
Sikkim	• All the eligible consumers can install RTS systems in the 1–500 kW range under net metering mechanism.
Tamil Nadu	• Net metering is available for domestic consumers from 1 kW to 999 kW. • Gross metering is available for all categories (except low tension category) from 151 kW to 999 kW. • Net billing or net feed-in is available for all categories (except hut and agriculture) up to the level of sanctioned load / contracted demand of their service connection or 999 kW whichever is lower.
Telangana	• Net metering available for all consumers from 1 kW to 1 MW. • Residential and government consumers can install up to 100% of their sanctioned load. • Industrial, commercial and other consumers can install up to 80% of their sanctioned load.
Tripura	• All eligible consumers can install RTS system within the capacity range of 1 kWp–1 MWp under net metering mechanism.

State	Applicable Metering Framework
Uttar Pradesh	• The maximum capacity of the system to be installed by any eligible consumer shall not exceed 100% of the sanctioned load/ connected load / contracted demand of the consumer. • The capacity of the grid- connected RTS system to be installed by any eligible consumer or third-party owner shall be in 1 kWp–2 MWp range. • Agriculture and domestic consumers are eligible for net metering. • All consumers are eligible for gross metering.
Uttarakhand	• Installed capacity at any eligible consumer's premises shall be up to a maximum of 80% of the consumer's sanctioned load/contract demand. • In case of domestic consumer, such installed capacity shall be irrespective of consumer's sanctioned load/contract demand. • The maximum installed capacity of rooftop solar (RTS) and small solar photovoltaic shall not be more than 1MW.
West Bengal	• The minimum size of the system that can be set up under net metering arrangement and net billing arrangement would be 1 kW and shall not exceed the sanctioned load (in kW) or the contract demand (in kilovolt ampere [kVA]). • Domestic consumers from 1 kW up to 5 kW are eligible for net metering.

Notes:

1) Wherever the word "system" is used, it means "rooftop solar photovoltaic system", unless specified otherwise.

2) The above mentioned applicable metering framework status is as on 30 June 2022. The respective website of each State Electricity Regulatory Commission may be visited to check for prevalent regulatory provisions and the applicability of the metering framework.

Source: Study team analysis.

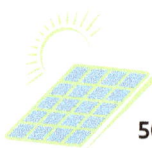